AF576001

Point-of-Care Detection Devices for Healthcare

Point-of-Care Detection Devices for Healthcare

Editor

Chao-Min Cheng

MDPI • Basel • Beijing • Wuhan • Barcelona • Belgrade • Manchester • Tokyo • Cluj • Tianjin

Editor
Chao-Min Cheng
National Tsing Hua University
Taiwan

Editorial Office
MDPI
St. Alban-Anlage 66
4052 Basel, Switzerland

This is a reprint of articles from the Special Issue published online in the open access journal *Diagnostics* (ISSN 2075-4418) (available at: https://www.mdpi.com/journal/diagnostics/special_issues/Paper_Based_Device).

For citation purposes, cite each article independently as indicated on the article page online and as indicated below:

LastName, A.A.; LastName, B.B.; LastName, C.C. Article Title. *Journal Name* **Year**, *Article Number*, Page Range.

ISBN 978-3-03943-659-0 (Hbk)
ISBN 978-3-03943-660-6 (PDF)

Cover image courtesy of Chao-Min Cheng.

Contents

About the Editor

Chao-Min Cheng received his Ph.D. in Biomedical Engineering in 2009 from Carnegie Mellon University. He then did his post-doctoral training at Harvard University. He is currently an independent P.I., tenured professor, at National Tsing Hua University, Taiwan, where he started in the summer of 2011. He has been blessed to receive the Outstanding Research Award from the Ministry of Science and Technology in Taiwan. He is currently an Editorial Board Member for Sensor Letters, Journal of Cellular and Molecular Medicine, Scientific Reports and Diagnostics. In addition to his academic contributions, Dr. Cheng serves as a consultant for biotech-based companies around the world with several Taiwan, China and U.S. patents granted.

Editorial

Integration of Mobile Devices and Point-Of-Care Diagnostic Devices—The Case of C-Reactive Protein Diagnosis

Xin-Fang Wu [1], Ching-Fen Shen [2,*] and Chao-Min Cheng [1,*]

1 Institute of Biomedical Engineering, National Tsing Hua University, Hsinchu 300, Taiwan; xinfangwu.tina@gmail.com

2 Department of Pediatrics, National Cheng Kung University Hospital, College of Medicine, National Cheng Kung University, Tainan 704, Taiwan

* Correspondence: drshen1112@gmail.com (C.-F.S.); chaomin@mx.nthu.edu.tw (C.-M.C.)

Received: 28 October 2019; Accepted: 6 November 2019; Published: 8 November 2019

In recent years, the misuse and overuse of antibiotics has promoted antibiotic resistance, which has now become a global public health concern. Among all prescribers, family practitioners are responsible for providing the majority of antibiotics prescriptions. Patterns of misuse are all related to each other and include not completing the full dosage of antibiotics provided, storing antibiotics for future use, sharing antibiotics, and taking antibiotics without a prescription. Not taking the antibiotics as prescribed not only wastes valuable medical resources but also exacerbates the spread of antibiotic resistance by strengthening bacterial strains. Appropriate antibiotic prescription practices and uses are essential for reducing the growing trend toward antibiotic resistance.

Recent research conducted in the United Kingdom by Christopher C. Butler, et al. [1] from January 2015 to February 2017, showed that point-of-care (POC) testing of C-reactive protein (CRP) used to assess acute exacerbations of chronic obstructive pulmonary disease (COPD) in primary care can safely reduce the use of antibiotics. This trial involved 653 patients from 86 general medical practices who were 40 years of age or older, had a diagnosis of COPD in their primary care clinical record, and were presented with an acute exacerbation of COPD with at least one of the Anthonisen criteria, including increased dyspnea, increased sputum volume, and the presence of purulent sputum. Before being randomly assigned into two groups, these patients were asked to complete a self-administered Clinical COPD Questionnaire to provide an assessment of COPD-related health, and the European Quality of Life-5 Dimensions-5 Level Questionnaire (EQ-5D-5L), measuring the adverse effects of antibiotics, health care utilization, and health utility. Randomization was used to divide the patients at a 1:1 ratio into one of two groups: (1) a CRP-guided group; and, (2) a usual-care group. Clinicians performed a CRP POC test as part of the CRP-guided group patient assessment and prescribed antibiotics based on the official guidance and their interpretation of CRP test results. The usual-care group did complete a CRP POC test. All patients underwent follow-up via telephone calls and face-to-face consultations. After 6 months, patients were asked to complete a Chronic Respiratory Disease Questionnaire (CRQ-SAS) and an EQ-5D-EL to determine disease-specific, health-related quality of health. Results indicate that fewer patients in the CRP-guided group reported antibiotic use, and fewer received antibiotic prescriptions at their initial consultation. In terms of general health status, the CRP-guided group had a higher health state score than the usual-care group, and the adjusted mean CRQ-SAS test difference between groups, which was small, indicated that there was no worsening of COPD-related health status. These pieces of evidence suggest that including a CRP POC test as part of the assessment for exacerbation of COPD in primary care may reduce patient-reported use of antibiotics as well as the prescribing of antibiotics by clinicians.

We believe that this research highlights the importance for providing more efficient medically based information to assist clinicians in gaining insight into exact patient health conditions. We recognize that whole blood tests in hospital settings are relatively costly and time-consuming, however, economical POC testing options exist that might facilitate more efficient initial examinations and more productive ultimate outcomes [2,3]. Our research group, for example, has focused on the development of an economical, paper-based CRP test that can be used in combination with a smartphone-based application to provide rapid and easily assessed CRP levels from whole blood (as shown in Figure 1).

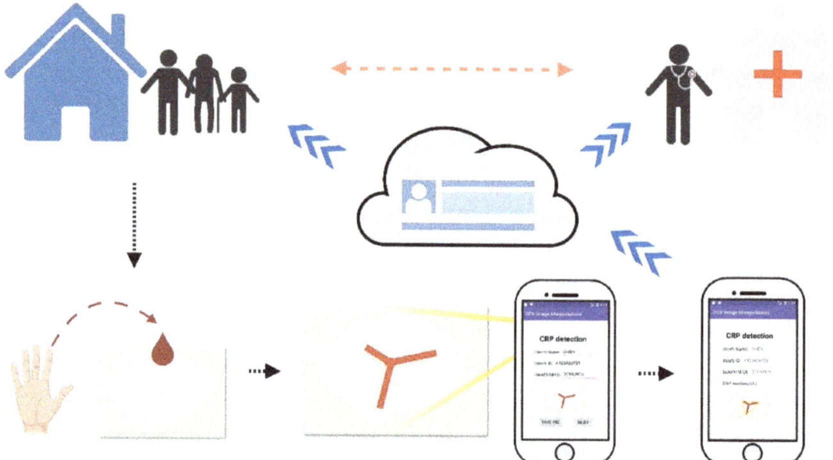

Figure 1. Schematic of the integration of point-of-care diagnostic tools and mobile devices. This figure presents the conceptual view of the clinical data flow of the point-of-care diagnostic tools and mobile devices in case of the C-reactive protein diagnosis that we have approached. Once the users place a drop of whole blood onto the diagnostic device, the whole blood would flow through the three channels. The users then could use the mobile application (APP) we have developed to take a picture of the diagnostic device with the final diagnostic result. The mobile application could help users both record and analyze the length difference of the diagnostic device, automatically compare with the guideline, tell the users the result and store the result in the cloud storage through the internet. In the cloud, the diagnostic result could be shared with the clinicians, enabling the clinicians to immediately gain insights about the patient situations without the patients needing to be near or even in the hospital.

Our study has a two-fold aim. We have already developed a new format for producing paper-based CRP detection devices based on the length difference between the control and test paper-based channels (versus the conventional immunoassay approach). This new CRP detection device uses an inflammation-based detection approach. It leverages a latex agglutination reaction in response to C-reactive proteins to produce a quantifiable response (stain length) that can be observed with the naked eye. By comparing the difference in stain lengths between the two channels of our device, we can determine the level of C-reactive proteins in whole blood. This paper-based CRP detection device is fabricated via a wax printing method that defines hydrophobic boundaries within a filter paper substrate. Our process reduces manufacturing and assay costs, speeds up operation, and can be used to purify and chromatographically correct the interference caused by whole-blood components using only a tiny sample of whole blood (only 5 μL) by relying on the hydrophilicity of filter paper [4]. Secondly, we have developed a smartphone-based application that allows users to both record and analyze the length differences between channels of our device, calculate the CRP level, compare results with the official guidelines, and then display the final diagnostic results to the end-user. By taking CRP diagnosis capacity away from clinical setting requirements and putting the first analytical step into the hands of home users and first-responders, patients and family practitioners are provided a better understanding of conditions and a better road map for improved health.

In an era in which smartphone use is ubiquitous, using mobile devices for disease diagnosis, prevention, and management is a promising and foreseeable future. Integration between mobile devices and POC tools could empower patients and provide rapid and accurate decision-making evidence for efficient diagnosis and subsequent care, especially in regards to the misuse or overuse of antibiotics. We believe that the study on the integration of the point-of-care diagnostic devices with a smartphone application (mobile devices) can be applied to facilitate a wide range of potential applications in POC diagnostics and handheld detection device development.

Funding: This research received no external funding.

Conflicts of Interest: The authors declare no conflict of interest.

References

1. Butler, C.C.; Gillespie, D.; White, P.; Bates, J.; Lowe, R.; Thomas-Jones, E.; Wootton, M.; Hood, K.; Phillips, R.; Melbye, H. C-reactive protein testing to guide antibiotic prescribing for COPD exacerbations. *N. Engl. J. Med.* **2019**, *381*, 1111–1120.
2. Gubala, V.; Harris, L.F.; Ricco, A.J.; Tan, M.X.; Williams, D.E. Point of care diagnostics: Status and future. *Anal. Chem.* **2011**, *84*, 487–515.
3. Cheng, C.-M.; Martinez, A.W.; Gong, J.; Mace, C.R.; Phillips, S.T.; Carrilho, E.; Mirica, K.A.; Whitesides, G.M. Paper-based ELISA. *Angew. Chem. Int. Ed.* **2010**, *49*, 47714–47774.
4. Lin, S.-C.; Tzeng, C.-Y.; Lai, P.-L.; Hsu, M.-Y.; Chu, H.-Y.; Tseng, F.-G.; Cheng, C.-M. Paper-based CRP monitoring devices. *Sci. Rep.* **2016**, *6*, 38171.

Communication

Paper-Based Detection Device for Alzheimer's Disease—Detecting β-amyloid Peptides (1–42) in Human Plasma

Wei-Hsuan Sung [1,2], Jung-Tung Hung [3], Yu-Jen Lu [2,4,*] and Chao-Min Cheng [5,*]

1 Chang Gung Memorial Hospital Linkou Medical Center, Taoyuan 33305, Taiwan; w.h.sung0109@gmail.com
2 College of Medicine, Chang Gung University, Taoyuan 33302, Taiwan
3 Institute of Stem Cell & Translational Cancer Research, Chang Gung Memorial Hospital Linkuo Medical Center, Taoyuan 33305, Taiwan; felixhjt@gmail.com
4 Department of Neurosurgery, Chang Gung Memorial Hospital Linkou Medical Center, Taoyuan 33305, Taiwan
5 Institute of Biomedical Engineering, National Tsing Hua University, Hsinchu 30013, Taiwan
* Correspondence: alexlu0416@gmail.com (Y.-J.L.); chaomin@mx.nthu.edu.tw (C.-M.C.)

Received: 12 April 2020; Accepted: 28 April 2020; Published: 30 April 2020

Abstract: The diagnosis of Alzheimer's disease (AD) is frequently missed or delayed in clinical practice. To remedy this situation, we developed a screening, paper-based (P-ELISA) platform to detect β-amyloid peptide 1–42 (Aβ42) and provide rapid results using a small volume, easily accessible plasma sample instead of cerebrospinal fluid. The protocol outlined herein only requires 3 μL of sample per well and a short operating time (i.e., only 90 min). The detection limit of Aβ42 is 63.04 pg/mL in a buffer system. This P-ELISA-based approach can be used for early, preclinical stage AD screening, including screening for amnestic mild cognitive impairment (MCI) due to AD. It may also be used for treatment and stage monitoring purposes. The implementation of this approach may provide tremendous impact for an afflicted population and may well prompt additional and expanded efforts in both academic and commercial communities.

Keywords: Alzheimer's disease; β-amyloid peptide; paper-based ELISA; P-ELISA, point of care testing

---

Alzheimer's disease (AD) is one the most common irreversible neurodegenerative diseases across the globe. The massive number of people affected worldwide totals nearly 44 million [1]. AD results in drastically impaired cognitive function and a reduced capacity to perform even daily routines and activities. Currently, AD diagnosis relies heavily on symptomology with symptom-dependent tools including guidance from the following: (a) National Institute of Neurological and Communicative Disorders and Stroke AD and Related Disorders Association (NINCDS-ADRDA, UK) and (b) Diagnostic and Statistical Manual of the American Psychiatric Association (DSM-IV/DSM-5) [2]. As a result, AD diagnosis is frequently missed or delayed in clinical practice [3]. More recent criteria such as those provided by the National Institute on Aging and the Alzheimer's Association (NIA-AA) include the use of biomarkers (e.g., β-amyloid and tau) for diagnostic support [4]. As a result, focus has rightly begun to shift toward developing early-stage methods for the detection of possibly potent AD biomarkers.

Most existing diagnostic methods, e.g., neuroimaging, enzyme-linked immunosorbent assay (ELISA), and polymerase chain reaction (PCR), are not suitable for point-of-care (POC) testing in their current state because they rely on highly sophisticated machinery and equipment, complicated operating procedures, and invasive or destructive sampling methods. Several newer studies have demonstrated greater creativity and have overcome problems by developing new POC testing devices to detect AD-related biomarkers. For example, an electrochemical immuno-sensing approach has been demonstrated for the detection of β-amyloid peptide 1–42 (Aβ42) at pM levels in a relatively shorter period of time than can be accomplished with an ELISA [5]. Stravalaci et al. described a novel immunoassay based on surface plasmon resonance (SPR) that specifically recognizes biologically active oligomers of the β-amyloid peptide (Aβ) [6]. Despite these advances, there is still an urgent need for rapid, effective, and easily used POC devices for early AD screening. The above-mentioned biosensors are not currently practical enough for clinical validation because they may be costly, involve a relatively time-consuming processes (e.g., immunoassay based on SPR requires a 5 h incubation period to produce a maximal signal), or they may require sophisticated signal readers. On the other hand, our paper-based POC device for the detection of Aβ42 is rapid, effective, inexpensive, and requires no sophisticated laboratory equipment. This process relies on an easily accessible body fluid, plasma, that facilitates minimally invasive first-step screening within one and a half hours.

A paper-based ELISA (P-ELISA) has previously been used to successfully detect proteins such as vascular endothelial growth factor (VEGF), as well as noncollagenous 16A (NC16A) autoimmune antibody toward diagnosis of various diseases such as age-related macular degeneration, bullous pemphigoid and *Escherichia coli* O157:H7 infection [7–10]. We have now demonstrated a P-ELISA system to detect Aβ42 in plasma. The aim of our study was twofold: (1) to expand the field of biomarker-dependent AD screening, as the use of biomarkers to support diagnosis has gained value and momentum, and, (2) to develop a specific POC tool using a P-ELISA to detect Aβ42 in both buffer and plasma systems. Based on its appropriate limit of detection (LOD), shorter operation duration, and lower cost, this method might set an example for the development of other approaches employing AD-related biomarkers for early stage screening, pre-treatment monitoring, in-treatment monitoring, and post-treatment follow-up. To our knowledge, our study is the first to apply a P-ELISA to detect plasma Aβ42.

Several studies have supported the important role of Aβ42 in the development of AD and have indicated that Aβ42 level dysregulation is responsible for the abnormal accumulation of Aβ42 plaques in the hippocampus and cortex [11,12]. For this reason, Aβ42 has been identified as a diagnostic biomarker, and anti-Aβ-directed therapies have been developed to combat AD [13]. With reliable detection at the core of any diagnostic approach, we first developed a buffer system-based P-ELISA tool to detect Aβ42 in 10-fold dilutions from 1 ng/mL to 1 pg/mL. An outline of our process is provided in Scheme 1 (below). After completing our P-ELISA process (as shown in the supporting movie), we visually interpreted the colorimetric output signal and used a smartphone camera (Apple, 1 Infinite Loop Cupertino, CA 95014, USA) to record the results. This process eliminates the need for any other specialized detector device. Colorimetric assays are particularly well-suited for use in resource-poor settings where plate readers and fluorescence scanners are rare but smartphones are relatively common. We converted our P-ELISA colorimetric results to eight-bit grayscale with ImageJ software using the formula: gray = (red + green + blue)/3. The color intensity was measured from min to max and defined as [experiment zone intensity] − [blank zone intensity]. The Mann–Whitney U test was used to compare the median mean intensity of different Aβ42 concentrations. The LOD was calculated as 63.04 pg/mL, as determined by nonlinear regression fits. Figure 1 displays the significant difference ($p < 0.001$) found between the group with concentrations at 1 ng/mL and our negative control group. The grayscale color intensity values at Aβ42 concentrations of 100, 10, and 1 pg/mL were significantly different ($p < 0.01$) compared to the grayscale color intensity value of the control group.

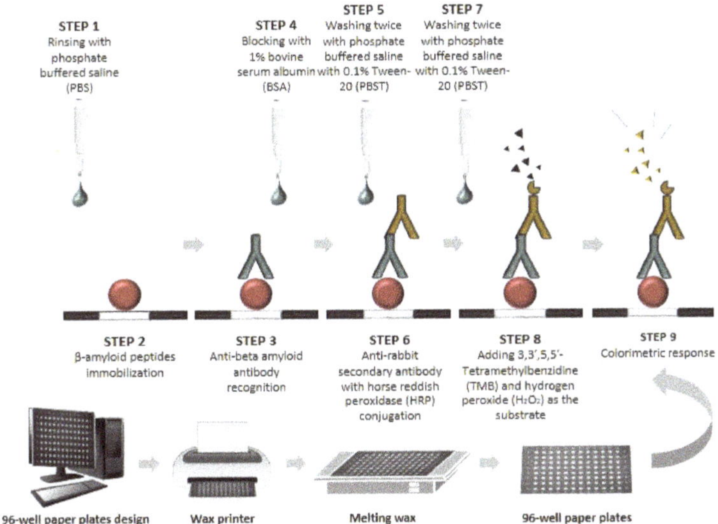

Scheme 1. Schematic of our paper-based ELISA (P-ELISA) device development and test procedure for the detection of β-amyloid peptide 1-42 (Aβ42) concentrations in both buffer and plasma systems.

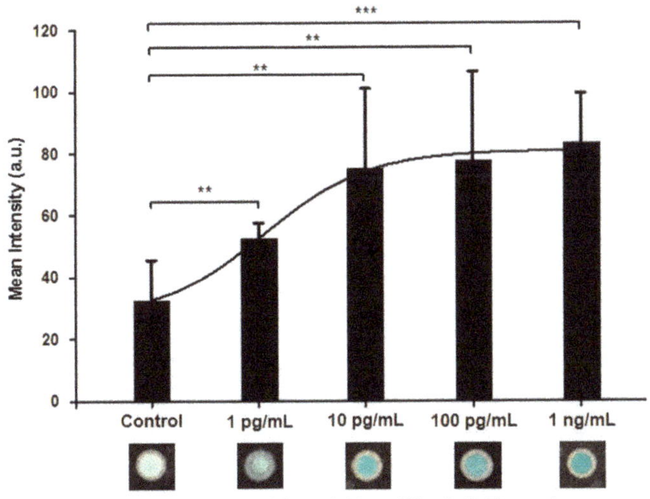

Figure 1. Colorimetric results (intensity) from our paper-based ELISA (P-ELISA) test for β-amyloid peptide 1-42 (Aβ42) concentrations in a buffer system. The color intensity difference between our 1 pg/mL Aβ42 concentration and our control was very significant. (** $p < 0.01$; *** $p < 0.001$).

Clinically, biomarkers have been used to screen for AD, but these approaches have required semi-invasive cerebrospinal fluid (CSF) sampling via lumbar puncture and/or the use of costly neuroimaging techniques [14]. Transitioning the use of these biomarkers to portable and reliable POC diagnostic devices has been challenging. Cerebrospinal fluid Aβ42 assays may be a more accurate reflection of the central amyloid pathology associated with AD, but there has been some reluctance to employ this approach for routine analysis because of the risk associated with external drains and severe disturbances in CSF [15]. For this reason and others, there have been increased interest and research into the use of more easily accessible sample sources, such as plasma, that contain measurable quantities of Aβ42 suitable for clinical assessment [16]. Previous studies have reported that intra-cerebroventricular injection of Aβ42 is correlated with plasma Aβ42 levels in a mouse model, thus confirming the in vivo mixing of CSF and plasma Aβ42 pools [17]. In humans, a weak positive correlation was also observed between plasma and CSF Aβ42 levels [18]. Moreover, increasing evidence had indicated that plasma Aβ42 concentration may be a risk predictor for AD [19], though some studies have produced controversial results [20]. Kim et al. outlined a filtration-based approach for distinguishing between normal plasma Aβ42 levels and those of patients with AD [21]. Mayeux et al. found mean plasma Aβ42 levels of 82.4 6 ± 8.6 pg/mL among patients with AD and subsequently found baseline mean plasma Aβ42 levels of 68.7 pg/mL and follow-up levels of 76.5 pg/mL in individuals with AD in a later study [22,23]. Using variable capture antibodies and analytical platforms, a wide range of mean plasma Aβ42 levels, from 36 to 140 pg/mL, have been reported in patients suffering from AD [24]. We elected to examine plasma Aβ42 concentration using our own unique P-ELISA approach, employing the same process and equipment employed in our buffer system analysis. We used four sets of plasma samples containing four different concentrations of Aβ42; 0 (control), 10 pg/mL, 100 pg/mL, and 1 ng/mL. For our secondary antibody, we used horseradish peroxidase (HRP) conjugated anti-rabbit antibody (Cat. No.: 7074, Cell Signaling Technology, 3 Trask Lane, Danvers, MA01923, USA) on plasma samples 1 and 2, and we used HRP-conjugated anti-rabbit antibody (Cat. No.: Ab6702, Abcam, Discovery Drive Cambridge Biomedical Campus, Cambridge CB2 0AX, UK) on plasma samples 3 and 4. A comparison between the two secondary antibodies is shown in Table 1. In Figure 2, plasma samples 1 and 2 displayed significant differences ($p < 0.05$) compared to the control for spiked Aβ42 concentrations of 100 pg/mL and 1 ng/mL, respectively. Plasma samples 3 and 4, however, displayed significant differences ($p < 0.05$) compared to the control for spiked Aβ42 concentrations of 10 and 100 pg/mL. From these results, we gathered that secondary antibody selection does appear to affect the performance of our P-ELISA platform. Our plasma system results were also approximately 10 times less sensitive than those from our buffer system. This may be explained by the fact that Aβ42 has to be measured in the matrix as a derivative of blood, which contains very high levels of plasma proteins such as albumin, clotting factor, and immunoglobulin G (IgG), all of which interfere with the application and interpretation of biochemical marker assay results [25,26]. There is room for improvement in the sensitivity and reliability for a plasma-based P-ELISA. Despite these difficulties, a plasma-based P-ELISA system may be used for early AD screening, as suggested by Blennow et al. [27]. Furthermore, repeated longitudinal measurements of plasma Aβ42 level may be useful for routine follow-up to determine disease progression and monitor therapy.

Table 1. Comparison between the two secondary antibodies used in our paper-based ELISA (P-ELISA) system for the detection of β-amyloid peptide 1–42 (Aβ42).

	Goat Anti-Rabbit IgG H and L (Cat. No.: Ab6702)	**Anti-Rabbit IgG, HRP-Linked Antibody (Cat. No.: 7074)**
Host Species	Goat	Goat
Target Species	Rabbit	Rabbit
Clonality	Polyclonal	Polyclonal
Isotype	IgG	IgG
Performance	10 pg/mL	100 pg/mL
Brand	Abcam	Cell Signaling Technology

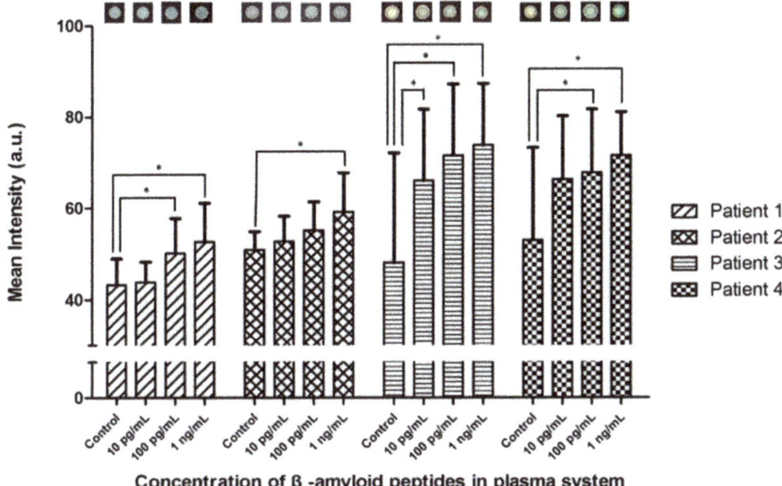

Figure 2. Colorimetric results (intensity) from a paper-based ELISA (P-ELISA) test for β-amyloid peptide 1–42 (Aβ42) concentration in a plasma system. The secondary antibody used in plasma 1 and 2 was from Cell Signaling Technology, while that used in plasma 3 and 4 was from Abcam. The limit of detection (LOD) for tests using plasma 1 and 2 was approximately 100 pg/mL, while the LOD for tests using plasma 3 and 4 was about 10 pg/mL. (* $p < 0.05$).

Clinical AD is thought to be preceded by a long asymptomatic or mildly symptomatic period that may be initiated 15–20 years Fprior to the onset of clinical signs [28]. This pre-dementia period is primarily composed of two parts: (1) preclinical AD and (2) amnestic mild cognitive impairment (MCI) due to AD development (Figure 3). Preclinical AD consists of the following three stages: (1) stage 1, which is manifested by the evidence of amyloidosis; (2) stage 2, which is characterized by not only amyloidosis but also evidence of neurodegeneration; an, (3) stage 3, a combination of amyloidosis, neurodegeneration, and subtle cognitive decline not meeting the criteria for MCI [29]. Compared to preclinical AD, amnestic MCI due to AD is defined as noticeable cognitive impairment resulting from underlying AD pathology. Because the development of AD is irreversible and progressive, there is an increasing need for biomarker-based screening tools to identify patients in preclinical or early clinical stages of AD. These patients would be greatly benefited by early intervention before more severe and irreversible damage occurs to the brain. In the past decade, a number of studies have made great efforts to develop biomarker-based screening tools and POC testing platforms to diagnose AD. Nakamura et al. validated the clinical utility of a blood-based Aβ assay using immunoprecipitation and mass spectrometry to predict brain Aβ burden [30]. Garyfallou et al. demonstrated an electrochemical immunosensor that can be easily integrated into portable devices to diagnose AD using plasma immunoglobulins [31]. Tonello et al. developed a POC testing system based on screen-printed electrochemical sensors (SPES) [32]. This study, however, is the first to demonstrate a P-ELISA system for Aβ42 detection in human plasma. It is challenging to measure Aβ42 due to antibody masking, Ab oligomerization, and Ab complex formation [33]. Plasma Aβ42 is also hard to use for diagnosing late-onset AD as a single time-point measure due to the considerable overlap with changes in the normal, aging population and the onset of vascular diseases [18,34]. We hope to promote the use of a P-ELISA for early screening, routine follow-up analyses, as well as AD monitoring in living patients as an adjunct to care. If detected at concentrations associated with risk, Aβ42 levels can be modified, as demonstrated by Boada et al., who describe a process for modifying Aβ42 concentration in plasma using plasma exchange (PE) and albumin replacement that improved cognition in patients with mild-to-moderate AD [35]. Our P-ELISA platform can help optimize therapeutics and improve disease progression prediction [36]. P-ELISA methods provide several advantages compared to conventional ELISA methods (Table 2). First, the entire P-ELISA process, from antigen immobilization to colorimetric reaction, can be completed within one and half hours; by contrast, a conventional ELISA requires at least six-to-eight hours to complete. Second, a P-ELISA requires only 3 μL of sample per test zone, while conventional ELISA requires more than twenty-five times that. Finally, P-ELISA results can be quantified with simple devices, such as smartphone cameras, which increases their usability and broadens their impact. Further research could result in the production of a paper-based multiplexed assay incorporating peptide-detecting ELISA to create a multi-step, all-in-one diagnostic device [37,38]. We have accomplished the first step toward this goal, creating a paper-based device for peptide detection, with this study.

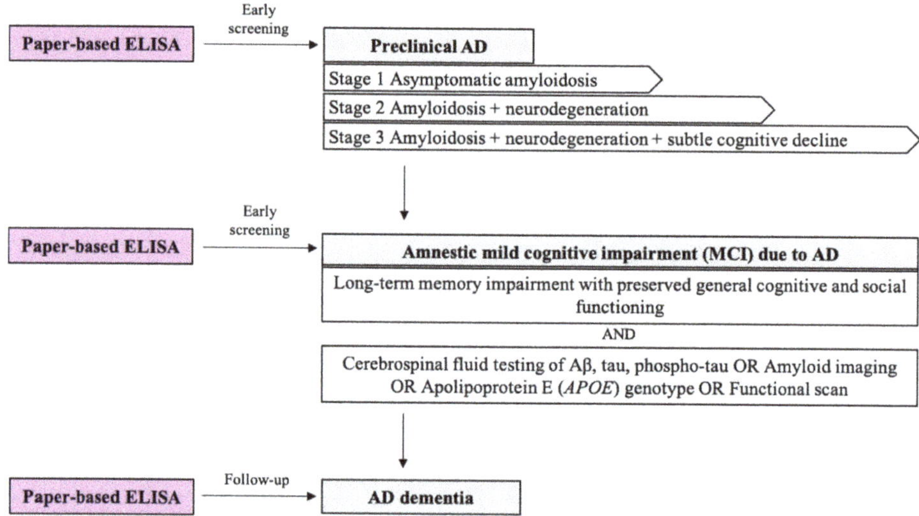

Figure 3. The role of point-of-care (POC) β-amyloid peptide 1–42 (Aβ42) testing for patients with preclinical Alzheimer's disease (AD), amnestic mild cognitive impairment (MCI) due to AD, and AD dementia.

Table 2. Comparison between the paper-based ELISA (P-ELISA) and conventional enzyme-linked immunosorbent assay (ELISA) systems for the detection of β-amyloid peptide 1–42 (Aβ42) using plasma and cerebrospinal fluid (CSF) samples.

	Paper-Based ELISA (P-ELISA)	Enzyme-Linked Immunosorbent Assay (ELISA) [25,39]	
Time	1.5 h	6–8 h (at least)	
Sample Volume (per Test Zone)	3 μL	75 μL	100–370 μL
Sample Source	Buffer	Plasma	CSF
Limit of Detection	63.04 pg/mL	5.71 pg/mL	312 pg/mL

This study outlines our development of the first P-ELISA tool for Aβ42 detection with demonstrated potential for testing human plasma. Our findings underscore the potential for employing a P-ELISA for both pre-clinical AD screening and post-diagnosis treatment monitoring. Compared to commonly-used Aβ42 detection methods, the P-ELISA offers five principal advantages: (1) rapidity, (2) small sample and reagent volume requirements, (3) cost-effectiveness, (4) readily available equipment and materials, and (5) improved clinical safety due to the fact that required samples involve the appropriation of plasma as opposed to CSF via lumbar puncture. P-ELISA techniques require some improvement in accuracy, precision, and long-term stability to render them more commercially viable. However, we found our approach to be highly sensitive, as evidenced by the 63.04 pg/mL LOD value attained in our buffer system experiments. In conclusion, our P-ELISA system is a promising candidate for the early screening of AD pre-dementia period and the post-diagnostic monitoring of AD, especially in small laboratories and in developing countries where cost and convenience are more critical.

Author Contributions: Conceptualization, Y.-J.L. and C.-M.C.; methodology, C.-M.C.; validation, J.-T.H.; formal analysis, W.-H.S.; investigation, W.-H.S.; resources, C.-M.C.; data curation, Y.-J.L.; writing—original draft preparation, W.-H.S.; writing—review and editing, W.-H.S.; visualization, W.-H.S.; supervision, Y.-J.L. and C.-M.C.; project administration, Y.-J.L. and C.-M.C.; funding acquisition, Y.-J.L. All authors have read and agreed to the published version of the manuscript.

Funding: The study was supported by the project 'CMRPG3F0883' of Linkou Chang Gung Memorial Hospital, Taiwan and the project 'MOST 107-2628-E-007-001-MY3' as well as the project 'MOST-107-2314-B-182-020' of Ministry of Science and Technology, Taiwan.

Conflicts of Interest: The authors declare no conflict of interest.

Abbreviations

AD	Alzheimer's disease
P-ELISA	Paper-based ELISA
Aβ42	β-amyloid peptide 1–42
MCI	Mild cognitive impairment
NINCDS-ADRDA	National Institute of Neurological and Communicative Disorders and Stroke AD and Related Disorders Association
DSM	Diagnostic and Statistical Manual of the American Psychiatric Association
NIA-AA	National Institute on Aging and the Alzheimer's Association
ELISA	Enzyme-linked immunosorbent assay
PCR	Polymerase chain reaction
POC	Point-of-care
SPR	Surface plasmon resonance
Aβ	ββrface plpeptide
VEGF	Vascular endothelial growth factor
NC16A	Noncollagenous 16A
LOD	Limit of detection
CSF	Cerebrospinal fluid
HRP	Horseradish peroxidase
IgG	Immunoglobulin G
SPES	Screen-printed electrochemical sensors
PE	Plasma exchange
APOE	Apolipoprotein E

References

1. Alzheimer's Association. 2018 Alzheimer's disease facts and figures. *Alzheimer's Dement.* **2018**, *14*, 367–429. [CrossRef]
2. Freedman, R.; Lewis, D.A.; Michels, R.; Pine, D.S.; Schultz, S.K.; Tamminga, C.A.; Gabbard, G.O.; Gau, S.S.; Javitt, D.C.; Oquendo, M.A.; et al. The initial field trials of DSM-5: New blooms and old thorns. *Am. J. Psychiatry* **2013**, *170*, 1–5. [CrossRef] [PubMed]
3. Bateman, R.J.; Xiong, C.; Benzinger, T.L.S.; Fagan, A.M.; Goate, A.; Fox, N.C.; Marcus, D.S.; Cairns, N.J.; Xie, X.; Tyler, M.S.; et al. Clinical and biomarker changes in dominantly inherited Alzheimer's disease. *N. Engl. J. Med.* **2012**, *367*, 795–804. [CrossRef]
4. Bature, F.; Guinn, B.A.; Pang, D.; Pappas, Y. Signs and symptoms preceding the diagnosis of Alzheimer's disease: A systematic scoping review of literature from 1937 to 2016. *BMJ Open* **2017**, *7*, e015746. [CrossRef] [PubMed]
5. Kaushik, A.; Jayant, R.D.; Tiwari, S.; Vashist, A.; Nair, M. Nano-biosensors to detect beta-amyloid for Alzheimer's disease management. *Biosens. Bioelectron.* **2016**, *80*, 273–287. [CrossRef] [PubMed]

6. Stravalaci, M.; Bastone, A.B.M.; Cagnotto, A.; Colombo, L.; Fede, D.G.; Tagliavini, F.; Cantu, L.; Del, F.E.; Mazzanti, M.C.R.; Salmona, M.; et al. Specific recognition of biologically active amyloid-beta oligomers by a new surface plasmon resonance-based immunoassay and an in vivo assay in Caenorhabditis elegans. *J. Biol. Chem.* **2012**, *287*, 27796–27805. [CrossRef] [PubMed]
7. Cheng, C.M.; Martinez, A.W.; Gong, J.; Mace, C.R.; Phillips, S.T.; Carrilho, E.; Mirica, K.A.; Whitesides, G.M. Paper-based ELISA. *Angew. Chem. Int. Ed. Engl.* **2010**, *49*, 4771–4774. [CrossRef]
8. Hsu, M.Y.; Hung, Y.C.; Hwang, D.K.; Lin, S.C.; Lin, K.H.; Wang, C.Y.; Choi, H.Y.; Wang, Y.P.; Cheng, C.M. Detection of aqueous VEGF concentrations before and after intravitreal injection of anti-VEGF antibody using low-volume sampling paper-based ELISA. *Sci. Rep.* **2016**, *6*, 34631. [CrossRef]
9. Pang, B.; Zhao, C.; Li, L.; Song, X.; Xu, K.; Wang, J.; Liu, Y.; Fu, K.; Bao, H.; Song, D.; et al. Development of a low-cost paper-based ELISA method for rapid Escherichia coli O157:H7 detection. *Anal. Biochem.* **2017**, *542*, 58–62. [CrossRef]
10. Hsu, C.K.; Huang, H.Y.; Chen, W.R.; Nishie, W.; Ujiie, H.; Natsuga, K.; Fan, S.T.; Wang, H.K.; Lee, J.Y.; Tsai, W.L.; et al. Paper-based ELISA for the detection of autoimmune antibodies in body fluid-the case of bullous pemphigoid. *Anal. Chem.* **2014**, *86*, 4605–4610. [CrossRef]
11. Barage, S.H.; Sonawane, K.D. Amyloid cascade hypothesis: Pathogenesis and therapeutic strategies in Alzheimer's disease. *Neuropeptides* **2015**, *52*, 1–18. [CrossRef]
12. Gouras, G.K.; Olsson, T.T.; Hansson, O. β-Amyloid peptides and amyloid plaques in Alzheimer's disease. *Neurotherapeutics* **2015**, *12*, 3–11. [CrossRef] [PubMed]
13. Lane, R.F.; Shineman, D.W.; Steele, J.W.; Lee LB, H.; Fillit, H.M. Beyond amyloid: The future of therapeutics for Alzheimer's disease. In *Advances in Pharmacology*; Academic Press: Cambridge, MA, USA, 2012; Volume 64, pp. 213–271.
14. Henriksen, K.; O'Bryant, S.E.; Hampel, H.; Trojanowski, J.Q.; Montine, T.J.; Jeromin, A.; Blennow, K.; Lonneborg, A.; Wyss-Coray, T.; Soares, H.; et al. The future of blood-based biomarkers for Alzheimer's disease. *Alzheimers Dement.* **2014**, *10*, 115–131. [CrossRef] [PubMed]
15. Laske, C.; Sohrabi, H.R.; Frost, S.M.; Lopez-de-Ipina, K.; Garrard, P.; Buscema, M.; Dauwels, J.; Soekadar, S.R.; Mueller, S.; Linnemann, C.; et al. Innovative diagnostic tools for early detection of Alzheimer's disease. *Alzheimer's Dement.* **2015**, *11*, 561–578. [CrossRef] [PubMed]
16. Poljak, A.; Crawford, J.D.; Smythe, G.F.; Brodaty, H.; Slavin, M.J.; Kochan, N.A.; Trollor, J.N.; Wen, W.; Mather, K.A.; Assareh, A.A.; et al. The relationship between plasma abeta levels, cognitive function and brain volumetrics: Sydney memory and ageing study. *Curr. Alzheimer Res.* **2016**, *13*, 243–255. [CrossRef]
17. Cho, S.M.; Kim, H.V.; Lee, S.; Kim, H.Y.; Kim, W.; Kim, T.S.; Kim, D.J.; Kim, Y. Correlations of amyloid-beta concentrations between CSF and plasma in acute Alzheimer mouse model. *Sci. Rep.* **2014**, *4*, 6777. [CrossRef]
18. Janelidze, S.; Stomrud, E.; Palmqvist, S.; Zetterberg, H.; van Westen, D.; Jeromin, A.; Song, L.; Hanlon, D.; Tan Hehir, C.A.; Baker, D.; et al. Plasma beta-amyloid in Alzheimer's disease and vascular disease. *Sci. Rep.* **2016**, *6*, 26801. [CrossRef]
19. Kawarabayashi, T.; Shoji, M. Plasma biomarkers of Alzheimer's disease. *Curr. Opin. Psychiatry* **2008**, *21*, 260–267. [CrossRef]
20. Lovheim, H.; Elgh, F.; Johansson, A.; Zetterberg, H.; Blennow, K.; Hallmans, G.; Eriksson, S. Plasma concentrations of free amyloid beta cannot predict the development of Alzheimer's disease. *Alzheimers Dement.* **2017**, *13*, 778–782. [CrossRef]
21. Kim, H.J.; Park, D.; Baek, S.Y.; Yang, S.-H.; Kim, Y.; Lim, S.M.; Kim, J.; Hwang, K.S. Dielectrophoresis-based filtration effect and detection of amyloid beta in plasma for Alzheimer's disease diagnosis. *Biosens. Bioelectron.* **2019**, *128*, 166–175. [CrossRef]
22. Mayeux, R.; Tang, M.-X.; Jacobs, D.M.; Manly, J.; Bell, K.; Merchant, C.; Small, S.A.; Stern, Y.; Wisniewski, H.M.; Mehta, P.D. Plasma amyloid β-peptide 1–42 and incipient Alzheimer's disease. *Ann. Neurol.* **1999**, *46*, 412–416. [CrossRef]
23. Mayeux, R.; Honig, L.S.; Tang, M.X.; Manly, J.; Stern, Y.; Schupf, N.; Mehta, P.D. Plasma Aβ40 and Aβ42 and Alzheimer's disease. *Neurology* **2003**, *61*, 1185. [CrossRef] [PubMed]
24. Toledo, J.B.; Shaw, L.M.; Trojanowski, J.Q. Plasma amyloid beta measurements—A desired but elusive Alzheimer's disease biomarker. *Alzheimer's Res. Ther.* **2013**, *5*, 8. [CrossRef] [PubMed]
25. Thambisetty, M.; Lovestone, S. Blood-based biomarkers of Alzheimer's disease: Challenging but feasible. *Biomark. Med.* **2010**, *4*, 65–79. [CrossRef]

26. Blennow, K.; Zetterberg, H. Understanding biomarkers of neurodegeneration: Ultrasensitive detection techniques pave the way for mechanistic understanding. *Nat. Med.* **2015**, *21*, 217–219. [CrossRef]
27. Blennow, K.; Zetterberg, H. Biomarkers for Alzheimer's disease: Current status and prospects for the future. *J. Intern. Med.* **2018**, *284*, 643–663. [CrossRef]
28. Mufson, E.J.; Ikonomovic, M.D.; Counts, S.E.; Perez, S.E.; Malek-Ahmadi, M.; Scheff, S.W.; Ginsberg, S.D. Molecular and cellular pathophysiology of preclinical Alzheimer's disease. *Behav. Brain Res.* **2016**, *311*, 54–69. [CrossRef]
29. Sperling, R.A.; Karlawish, J.; Johnson, K.A. Preclinical Alzheimer disease-the challenges ahead. *Nat. Rev. Neurol.* **2013**, *9*, 54–58. [CrossRef]
30. Nakamura, A.; Kaneko, N.; Villemagne, V.L.; Kato, T.; Doecke, J.; Doré, V.; Fowler, C.; Li, Q.-X.; Martins, R.; Rowe, C.; et al. High performance plasma amyloid-β biomarkers for Alzheimer's disease. *Nature* **2018**, *554*, 249–254. [CrossRef]
31. Garyfallou, G.Z.; Ketebu, O.; Sahin, S.; Mukaetova-Ladinska, E.B.; Catt, M.; Yu, E.H. Electrochemical Detection of Plasma Immunoglobulin as a Biomarker for Alzheimer's Disease. *Sensors* **2017**, *17*, 2464. [CrossRef]
32. Tonello, S.; Serpelloni, M.; Lopomo, N.F.; Abate, G.; Uberti, D.L.; Sardini, E. Screen-printed biosensors for the early detection of biomarkers related to alzheimer disease: Preliminary results. *Procedia Eng.* **2016**, *168*, 147–150. [CrossRef]
33. Galozzi, S.; Marcus, K.; Barkovits, K. Amyloid-beta as a biomarker for Alzheimer's disease: Quantification methods in body fluids. *Expert Rev. Proteom.* **2015**, *12*, 343–354. [CrossRef] [PubMed]
34. Poljak, A.; Sachdev, P.S. Plasma amyloid beta peptides: An Alzheimer's conundrum or a more accessible Alzheimer's biomarker? *Expert Rev. Neurother.* **2017**, *17*, 3–5. [CrossRef] [PubMed]
35. Boada, M.; Anaya, F.; Ortiz, P.; Olazaran, J.; Shua-Haim, J.R.; Obisesan, T.O.; Hernandez, I.; Munoz, J.; Buendia, M.; Alegret, M.; et al. Efficacy and safety of plasma exchange with 5% albumin to modify cerebrospinal fluid and plasma amyloid-beta concentrations and cognition outcomes in alzheimer's disease patients: A multicenter, randomized, controlled clinical trial. *J. Alzheimers Dis.* **2017**, *56*, 129–143. [CrossRef]
36. Snyder, H.M.; Carrillo, M.C.; Grodstein, F.; Henriksen, K.; Jeromin, A.; Lovestone, S.; Mielke, M.M.; O'Bryant, S.; Sarasa, M.; Sjøgren, M.; et al. Developing novel blood-based biomarkers for Alzheimer's disease. *Alzheimer's Dement.* **2014**, *10*, 109–114. [CrossRef]
37. Sher, M.; Zhuang, R.; Demirci, U.; Asghar, W. Paper-based analytical devices for clinical diagnosis: Recent advances in the fabrication techniques and sensing mechanisms. *Expert Rev. Mol. Diagn.* **2017**, *17*, 351–366. [CrossRef]
38. Yamada, K.; Shibata, H.; Suzuki, K.; Citterio, D. Toward practical application of paper-based microfluidics for medical diagnostics: State-of-the-art and challenges. *Lab Chip* **2017**, *17*, 1206–1249. [CrossRef]
39. Perez-Grijalba, V.; Fandos, N.; Canudas, J.; Insua, D.; Casabona, D.; Lacosta, A.M.; Montanes, M.; Pesini, P.; Sarasa, M. Validation of immunoassay-based tools for the comprehensive quantification of abeta40 and abeta42 peptides in Plasma. *J. Alzheimers Dis.* **2016**, *54*, 751–762. [CrossRef]

Article

Preliminary Assessment of Burn Depth by Paper-Based ELISA for the Detection of Angiogenin in Burn Blister Fluid—A Proof of Concept

Shin-Chen Pan [1,2], Yao-Hung Tsai [3], Chin-Chuan Chuang [3] and Chao-Min Cheng [3,*]

1 Department of Surgery, Section of Plastic and Reconstructive Surgery, National Cheng Kung University Hospital, College of Medicine, National Cheng Kung University, Tainan 704, Taiwan; pansc@mail.ncku.edu.tw

2 International Center for Wound Repair and Regeneration, National Cheng Kung University, Tainan 704, Taiwan

3 Institute of Biomedical Engineering and Frontier Research Center on Fundamental and Applied Sciences of Matters, National Tsing Hua University, No. 101, Sec. 2, Kuang-Fu Rd., East Dist., Hsinchu 300, Taiwan; michaeltsai45@gmail.com (Y.-H.T.); jason20011122@gmail.com (C.-C.C.)

* Correspondence: chaomin@mx.nthu.edu.tw; Tel.: +886-3516-2420

Received: 21 January 2020; Accepted: 25 February 2020; Published: 27 February 2020

Abstract: Rapid assessment of burn depth is important for burn wound management. Superficial partial-thickness burn (SPTB) wounds heal without scars, but deep partial-thickness burn (DPTB) wounds require a longer healing time and have a higher risk of scar formation. We previously found that DPTB blister fluid displayed a higher angiogenin level than SPTB blister fluid by conventional ELISA. In this study, we developed a paper-based ELISA (P-ELISA) technique for rapid assessment of angiogenin concentration in burn blister fluid. We collected six samples of SPTB blister fluid, six samples of DPTB blister fluid, and seven normal healthy serum samples for analysis. We again chose ELISA to measure and compare angiogenin levels across all of our samples, but we developed a P-ELISA tool and compared sample results from that tool to the results from conventional ELISA. As with conventional ELISA, DPTB blister fluid displayed higher angiogenin levels than SPTB in P-ELISA. Furthermore, our P-ELISA results showed a moderate correlation with conventional ELISA results. This new diagnostic technique facilitates rapid and convenient assessment of burn depth by evaluating a key molecule in burn blister fluid. It presents a novel and easy-to-learn approach that may be suitable for clinically determining burn depth with diagnostic precision.

Keywords: partial-thickness burn injury; burn blister fluid; P-ELISA; angiogenin; burn wound healing

1. Introduction

Burn wound prognosis depends on early and rapid diagnosis of burn depth. Superficial partial-thickness burn (SPTB) wounds heal spontaneously within two weeks of injury without scar formation. However, deep partial-thickness burn (DPTB) wounds take more than two weeks to heal and often result in hypertrophic scar formation if no aggressively surgical management. Optimal management of DPTB wounds can prevent skin scarring. Several methods were reported to assess the wound depth, including biopsy, thermography, laser Doppler techniques, and bedside clinical judgment [1]. Clinical observations are the gold standard for estimating clinical outcomes [2]. However, clinical assessment and prognosis of second-degree burn wounds with intact blisters become difficult in some cases, even for experienced surgeons. Measurements of tissue perfusion in injured wounds appear to be an option to assess tissue damage extent [3]. Although laser Doppler perfusion imaging was reported to be an efficient tool to evaluate blood flow of burn wounds [4–6], this machine is not readily available in all clinical settings.

Burn blisters, common to both SPTB and DPTB, are formed by an inflammatory response in early burn injury and exist between the epidermis and dermis [7]. The cytokines found in burn blister fluid can also be generated by activated or injured parenchymal cells after appropriate stimulation [8]. In our previous study, we observed that SPTB and DPTB blister fluids expressed different levels of angiogenin. When compared with SPTB, DPTB blister fluids displayed a higher angiogenin level [9]. In addition, angiogenin promoted in vitro angiogenesis such as endothelial cell proliferation and differentiation of circulating angiogenic cells as well as in vivo neovascularization [9]. Angiogenin, originally identified from a conditioned culture medium of colon cancer cells [10], is a potent angiogenic inducer and an independent prognostic factor in many cancers [11]. Because the angiogenin levels of SPTB and DPTB blister fluids were significantly different, the measurement of angiogenin in burn blister fluids can be used as a novel, noninvasive, or minimally invasive tool for surveying burn wound status.

ELISA is a well-established diagnostic tool for evaluating disease activity and assessing burn depth via analysis of fluid contents [9,12]. However, commercial ELISA kits are expensive, time-consuming, and are only available in clinical settings equipped with an ELISA reader. In order to simplify and expand clinical use of ELISA for early and efficient assessment of burn wounds, the development of a low-cost and rapid diagnostic tool for assessing burn depth is vital. Dot immunoassays on nitrocellular and filter papers is an established diagnostic tool [13–15]. Paper-based ELISA (P-ELISA), first developed by the Whitesides Research Group [16], has been an established diagnostic tool for decades. It is a useful procedure for performing immunoassays using a piece of filter paper for antibody–antigen recognition and has been applied for the diagnosis of infectious diseases (e.g., HIV and dengue fever), ophthalmological diseases (e.g., proliferative diabetic retinopathy, age-related macular degeneration), female genital diseases, and bullous pemphigoid [17–20]. The advantages of P-ELISA include speed, cost, small sample demands, and similar levels of sensitivity and specificity compared to conventional plate ELISA. Due to differential angiogenin levels between SPTB and DPTB blister fluids, the objective of this study is to develop a new application of the P-ELISA tool for the measurement of angiogenin expression in burn blister fluids in order to diagnose burn depth and facilitate improved burn wound care. Through conducting a small amount of clinical testing, we prove the clinical applicability of the device and guide the engineering process for further point-of-care and care-at-home devices with potential pre/clinical applications [21,22].

2. Material and Methods

2.1. Patient Samples

In order to study the differential expression of angiogenin in burn blisters from wounds of different depths, burn blister fluids were blindly aspirated with a needle from individually intact blisters within the first 3 days following injury and before identification of burn depth. Blister fluids were not classified into superficial or deep groups when fluids were harvested. Burn depth was confirmed by retrospective review of patient data according to healing status after the wound was healed. SPTB wounds were defined as those that healed within 14 days, and DPTB wounds were defined as those that required more than 14 days to heal or required burn wound debridement in the early assessment of wounds. Exclusion criteria included any potential bias such as fever, wound infection, or other severe medical problems including end-stage renal disease. Blood serum samples taken from healthy subjects were regarded as control samples. Informed consent was obtained from all patients, and study procedures were conducted in accordance with the Declaration of Helsinki and were approved by the Ethics Committee of National Cheng Kung University Hospital (No. A-ER-106-239, approval date 11 November 2017). All experiments were performed in accordance with relevant guidelines and regulations.

2.2. Preparation of Paper-Based 96-Well Plates Via Wax Printing

We designed a wax printing method for patterning Whatman No.1 filter papers (GE Healthcare, Buckinghamshire, UK) [20]. Briefly, a 96-well template was designed on a computer with Microsoft Office and then printed onto paper with wax. Complete, paper-depth hydrophobic barriers were created by melting the paper with printed-on wax at 105 °C for 5 min. Paper is highly permeable, which allows the wax to penetrate into the fiber matrix and then solidify to form defined hydrophobic barriers. This process is well-established and easy to complete.

2.3. Procedure of Paper-Based ELISA and Plate ELISA

Test zones were initially rinsed with 2 μL of Tris-buffered saline (TBS). Three microliters of burn fluid sample was then loaded onto the test zone and allowed to rest for 10 min. The test zone was then blocked with 3 μL 1% bovine serum albumin (BSA) for 10 min before adding 3 μL of rabbit anti-human angiogenin polyclonal primary antibody (Cat. No.: ab95389. Abcam, Cambridge, UK) and waiting another 10 min. Test zones were subsequently washed with 5 μL TBST (TBS+0.05% Tween 20), and then washed again with 10 μL TBST for a total of 2 washes. The washing process relied on a piece of blotting paper at the bottom to remove the washing buffer with capillary force. After washing, the test zone was incubated for 10 min with anti-rabbit IgG secondary antibody conjugated with horseradish peroxidase (HRP, Cat. No.: ab6721. Abcam, Cambridge, UK) before being washed again with 5 μL and 10 μL TBST (two washes). The color of the tested paper was developed after incubation with 3 μL of substrate solution (3,3′,5,5′-Tetramethylbenzidine (TMB): H_2O_2 = 1:1, Cat. No.: 555214. BD, Franklin Lakes, New Jersey, USA) for 10 min. The image signal was recorded using a camera at two time periods: (1) after 10 min of incubation with a secondary antibody; and (2) after 10 min of reaction time with the substrate solution. Both images were processed with Photocap software and analyzed with ImageJ software. To determine color intensity change, before and after images were first converted into 8-bit grayscale. Mean intensity was determined by comparing grayscale value differences from before and after images. Further normalization of the results was achieved by expressing values in terms of relative intensity, which was defined as (mean intensity of [experiment group]–mean intensity of [control group])/mean intensity of [control group]). Plate ELISA (Cat. No.: ab10600. Abcam, Cambridge, UK; Avastin® (bevacizumab), Roche, Basel, Switzerland) was used to measure the angiogenin and VEGF (Vascular endothelial growth factor) values in burn blister fluids. Each sample was applied in triplicate, and the average data from two plates were taken as final readouts. Values were expressed as mean ± SD. To also examine the role of pH on burn wound status, burn blister fluid pH values were collected with pH test strips (Cat. No.: 92111. Machery-Nagel, Düren, DE).

2.4. Data Analysis

Differences in angiogenin levels between SPTB and DPTB blister fluids was assessed using the Mann–Whitney U test. Analysis of variance was used to determine statistical differences among multiple groups. The relationship between the titer from the conventional plate ELISA and the relative intensity of P-ELISA was correlated using Pearson's correlation coefficient. Values of $p < 0.05$ were considered statistically significant. The results were expressed as mean ± SD.

3. Results

We have attempted to analyze the angiogenin levels in burn blister fluids to assess burn depth using P-ELISA. To determine optimal primary antibody concentration, we examined colorimetric responses to different dosages of recombinant angiogenin antibody on test paper impregnated with blister fluid. The results showed that 1 μg /mL of anti-angiogenin antibody provided the best signal. (Figure S1). To achieve the best staining with minimal background interference, we tested three different concentrations (0.05, 0.04, and 0.033 μg/mL) of secondary antibody. After testing serial dilutions of secondary antibody, we found that 0.033 μg/mL demonstrated the highest signal-to-background ratio for our primary antibody (Figure S2). We washed the test paper twice, first with 5 μL of TBST solution and then again with 10 μL of TBST solution, to remove unbound material. To determine optimal BSA concentration during the blocking step, we coated filter paper with different concentrations (1%, 0.5%, 0.1%) of blocking solution and compared reaction results to blister fluid harvested from DPTB patients and control blood serum. Mirroring the expected conventional plate ELISA results, our data showed that 1% of BSA displayed the best blocking effect for burn blister fluid when compared to our healthy human blood serum control (Figure S3). In addition to optimizing blocking, determining the best washing volume is essential for removing unbound reagents and reducing background signal. Unlike the conventional plate ELISA wash volume of 300 μL, P-ELISA requires only minute wash volumes. We tested the efficacy of three different wash solution amounts, 15 μL, 10 μL, and 5 μL, and found that three separate washes (the typical number of washes for conventional ELISA) with 5 μL removed all unbound nonspecific material (Figure S4). While the typical number of washes with conventional ELISA is three, we discovered that two cycles, one with 5 μL of wash buffer and one with 10 μL of wash buffer, provided the lowest background and strongest signal strength for P-ELISA (Figure S5).

Paper-based diagnostics provide a low-cost and easy-to-handle approach for studying a variety of target factors. Here, we demonstrated this approach by adding burn blister fluids and select reagents to test zones to detect angiogenin levels. As shown in Figure 1, burn fluid was added by hand, reagent was spotted onto each test zone, and the reaction was allowed to proceed for several minutes. Primary and secondary antibody were then added and optimized washing processes carried out to remove unbound antibody from the paper. Colorimetric responses were digitally recorded before and after color development. By analyzing the change in grayscale intensity of before and after images, we could determine the amount of angiogenin in burn blister fluids.

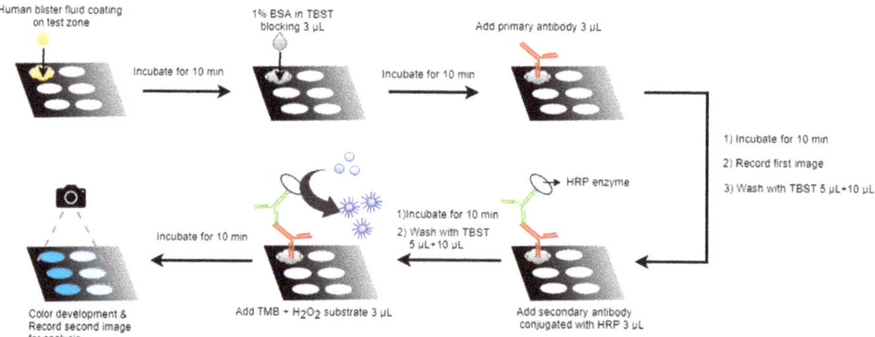

Figure 1. Procedure of paper-based ELISA (P-ELISA) for angiogenin analysis in burn blister fluid. Three microliters of burn fluid was loaded onto the test zone, and the test paper was incubated for 10 min. Three microliters of 1% BSA was then added for blocking, and the test paper was incubated for 10 min. The first image was recorded after loading 3 μL primary antibody, and the test paper was incubated for 10 min. The test paper was washed with 5 μL and then 10 μL TBST (Tris-buffered saline + 0.05% Tween 20) before adding 3 μL of horseradish peroxidase (HRP)-conjugated secondary antibody. The final procedure was to wash again with TBST and add 3 μL of substrate solution. After 10 min of incubation, we recorded the second image.

Relative mean intensity for angiogenin from SPTB burn blister fluids (Figure 2a) and DPTB burn blister fluids (Figure 2b) were tested and compared. In keeping with our previous study results, we found that angiogenin levels were significantly higher in DPTB fluids (4.2 ± 1.5, 95% confidence interval, 1.3–6.5, n = 6) than SPTB fluids (1.4 ± 0.3, 95% confidence interval, 0–4.0, n = 6) and healthy blood serum samples (0.9 ± 0.3, 95% confidence interval, 0.4–1.2, N = 7, $p < 0.01$) as measured by P-ELISA (Figure 2c). This demonstrates the reliability of P-ELISA to determine burn severity by evaluating angiogenin levels in burn blister fluids. Burn fluid samples have previously been tested via conventional ELISA plates for angiogenin and VEGF. Angiogenin is a downstream molecule of VEGF-regulated angiogenesis [23]. Observation of the role of VEGF in burn wound determination is interesting. Although no significant differences in VEGF were observed (SPTB: 31.5 ± 3.6 ng/mL, 95% confidence interval, 20.0–42.9 ng/mL, DPTB: 47.1 ± 10.9 ng/mL, 95% confidence interval, 16.8–77.5 ng/mL), there was a trend for higher angiogenin concentration in DPTB fluids compared with SPTB fluids as determined by conventional plate ELISA analysis (SPTB: 119.2 ± 28.4 ng/mL, 95% confidence interval, 28.8–209.6 ng/mL, DPTB: 331.5 ± 81.0 ng/mL, 95% confidence interval, 123.3–539.7 ng/mL, Figure 3, Table S1). No significant differences in VEGF level were observed between two different burn fluids, as shown in Figure 3, which was consistent with our previous study and another study indicating that VEGF was not responsible for differentiation of circulating angiogenic cells [9] or tumor growth and angiogenesis [23]. The consistent comparability of our P-ELISA results with conventional plate ELISA results was further supported by a correlation test using Pearson's correlation analysis. A moderate positive correlation (rho = 0.5906, $p = 0.0722$) between the titer of the plate ELISA and the relative intensity of P-ELISA was observed and is displayed in Figure 4.

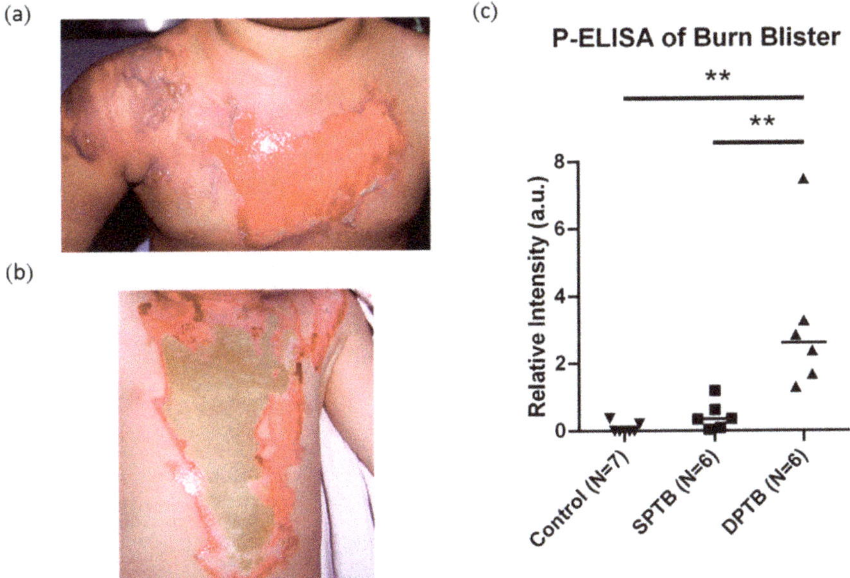

Figure 2. Analysis of angiogenin levels from superficial partial-thickness burn (SPTB) and deep partial-thickness burn (DPTB) blister fluids using P-ELISA. (**a**,**b**) Clinical pictures of superficial partial-thickness burn (SPTB, **a**) and deep partial-thickness burn (DPTB, **b**) wounds. (**c**) Comparison of angiogenin levels from two different burn fluids and healthy human blood serum as the control ($n = 6$ in SPTB and DPTB, $n = 7$ in control, mean ± S.D, ** $p < 0.01$).

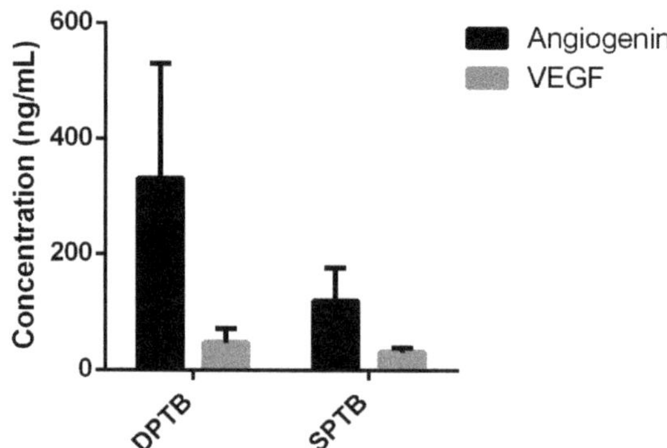

Figure 3. Analysis of angiogenin and VEGF (Vascular endothelial growth factor) concentrations in SPTB and DPTB blister fluids with conventional plate ELISA. A trend toward higher angiogenin concentration in DPTB fluids was detected, compared to SPTB ($n = 4$ in SPTB, $n = 6$ in DPTB, mean ± S.D.; $p = 0.07$). No significant difference in VEGF levels was observed between two different blister fluids ($n = 4$ in SPTB, $n = 5$ in DPTB, mean ± S.D.; $p = 0.26$).

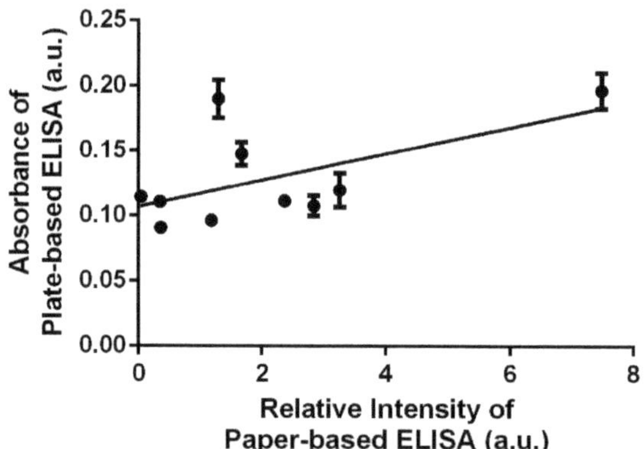

Figure 4. Correlation of the angiogenin detection between conventional plate ELISA and paper-based ELISA in burn blister fluids. The data show a moderate correlation between the results of P-ELISA and conventional plate ELISA ($r = 0.5906$, $p = 0.0722$).

In our laboratory, we have developed and explored the capacities for P-ELISA made of simple filter paper. To support the concept of P-ELISA material and process viability, we used patterned filter paper to collect directly absorbed blister fluid from human burn wounds (Supporting Movie) and had the test paper delivered to our laboratory on dry ice within 24 h, where we used it to successfully detect angiogenin levels via P-ELISA (Figure 5).

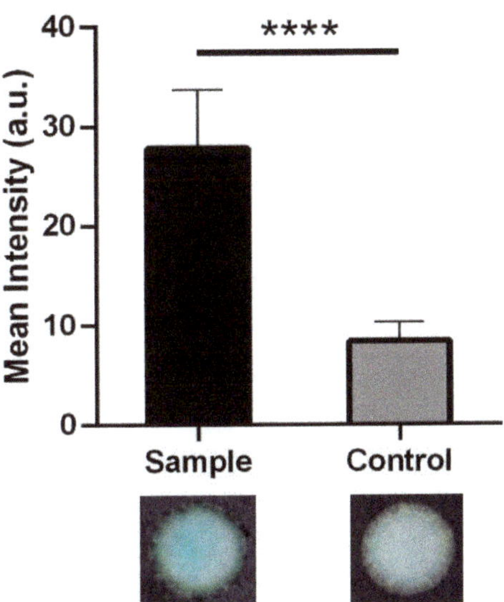

Figure 5. Clinical examination of angiogenin concentration by paper-based ELISA. Test paper was designed to absorb blister fluid from burn patients. Angiogenin signals were captured and analyzed. The mean intensity of detected angiogenin in burn blister fluid was significantly higher than that in normal serum control (mean ± S.D.; **** $p < 0.0001$, $n = 6$).

4. Discussion

Advances in diagnostic techniques have allowed clinicians to monitor disease severity in a rapid and noninvasive fashion. With regard to diagnosis of patients with ocular disease, P-ELISA provided a fast and sensitive VEGF assay of aqueous humor to monitor diseases such as senile cataracts, proliferative diabetic retinopathy, age-related macular degeneration, and retinal vein occlusion [17]. P-ELISA has also demonstrated the capacity for complex material collection capacity and multivalue measurement. It has, for instance, been used for outer macromolecule elimination and inner cervicovaginal fluid absorption to detect lactate concentration, glycogen concentration, and pH value in female genital diseases [19]. For patients with bullous pemphigoid, P-ELISA also delivered a simpler and faster diagnostic tool for detection of noncollagenous 16A (NC16A). By examining NC16A concentration with P-ELISA, bullous pemphigoid presence and disease state can be easily identified [20].

Previous concepts for burn depth assessment depend on measurements of tissue perfusion [24]. Laser Doppler imaging (LDI) studies are one of the most popular clinical techniques to assess burn depth. LDI is advantageous due to high accuracy and reduced invasiveness [25–27]. However, most clinical experiences using these modalities do not include examinations with intact blisters, which is a confounding factor that may skew analyzed results.

In this study, we sought to determine the burn depth by measuring angiogenin levels in burn blister fluid using the P-ELISA method. Evaluation of burn fluid contents by conventional ELISA has been successfully used to measure burn depth [9]. However, the necessity for longer operation duration severely hampered plate ELISA techniques for clinical practice. P-ELISA features operation process advantages compared to conventional plate ELISA. This P-ELISA technique required only 15 μL of reagent and an hour of processing compared to the 550 μL and 7 h–8 h required for conventional plate ELISA. Additionally, conventional plate ELISA requires a plate reader and P-ELISA results can be recorded with a camera (Table 1). Altogether, P-ELISA offers clear advantages for rapid and minimally invasive burn depth diagnosis and wound management. It may provide an easy and cost-effective method for any healthcare provider to assess burn wounds with excellent diagnostic precision and without the obstacle of a learning curve. However, P-ELISA is still conducted in a qualitative manner till now. Further refinement is needed to improve our device.

Table 1. Comparison of paper-based ELISA and conventional plate ELISA.

	Paper-Based ELISA		Conventional Plate ELISA	
antigen	angiogenin		angiogenin	
Primary antibody	Rabbit polyclonal anti-human angiogenin		Mouse monoclonal anti-human angiogenin	
Secondary antibody	Goat anti-rabbit lgG (HRP)		Goat anti-mouse lgG (HRP)	
Enzyme/substrate	HRP/TMB + H_2O_2		HRP/TMB + H_2O_2	
Detection device	camera		plate reader	
analysis	qualitative		quantitative	
Time and reagents	Volume (μL)	Time (min)	Volume (μL)	Time (min)
(1) antigen immobilization	3	10	50	120
(2) blocking	3	10	200	120
(3) primary antibody	3	10	100	120
(4) secondary antibody	3	10	100	120
(5) signal amplification	3	10	100	15
total	15	50	550	495

Wound pH may affect the healing process. A lower pH value was reported to be favorable for wound healing [28]. One study showed that healing burn wounds displayed a lower pH level (7.32) compared to unhealing burn wounds (pH 7.73) [29]. We observed an elevated alkaline pH value (8–9) in our burn fluid samples (Table S1). Aspiration of burn fluids in the early stages of injury may be responsible for this phenomenon. This finding is also consistent with a previous study showing that initial stages of healing generated a more basic pH compared to the relatively acidic environment of a repaired wound [30].

We feel that P-ELISA shows great promise for potential use in clinical practice. Although the clinical application of P-ELISA in burn wound assessment remains to be further studied, we look forward to seeing the application of this new technique in burn wounds. More cases are needed to verify our hypothesis prior to clinical adoption. Further optimization of the paper pattern and additional clinical trials will improve and advance the process and its implementation as a possible tool for healthcare.

5. Conclusions

We designed a less-invasive and faster method to assess burn depth using P-ELISA to determine angiogenin levels from burn blister fluid. This approach is cost-effective, easy to use, and potentially accurate. Despite a moderate correlation between conventional plate ELISA and P-ELISA, we demonstrated that P-ELISA is feasible to quantify blister angiogenin levels with distinct results to diagnose SPTB and DPTB wounds. We expect this novel medicine study to pave a possible path toward a simple diagnostic method for assessing burn depth without difficulty.

Supplementary Materials: The following are available online at http://www.mdpi.com/2075-4418/10/3/127/s1.

Author Contributions: Conceptualization, S.-C.P. and C.-M.C.; data curation, Y.-H.T. and C.-C.C.; formal analysis, Y.-H.T.; funding acquisition, S.-C.P. and C.-M.C.; investigation, Y.-H.T. and C.-C.C.; methodology, Y.-H.T.; project administration, C.-M.C.; resources, S.-C.P. and C.-M.C.; supervision, C.-M.C.; writing–original draft, S.-C.P.; Writing–review and editing, C.-M.C. All authors have read and agreed to the published version of the manuscript.

Funding: This study is funded by research grants from the Ministry of Science and Technology, Taiwan (MOST 107-2628-E-007-001-MY3 to C.-M.C.; MOST 108-2314-B-006-080-MY3 to S.-C.P.), National Cheng Kung University Hospital (NCKUH-10705018), the International Center for Wound Repair and Regeneration at National Cheng Kung University from The Featured Areas Research Center Program within the framework of the Higher Education Sprout Project by the Ministry of Education (MOE) in Taiwan, and by the grant of the Ministry of Science and Technology (MOST 107-3017-F-006-002). and the Frontier Research Center on Fundamental and

Applied Sciences of Matters from The Featured Areas Research Center Program within the framework of the Higher Education Sprout Project by the Ministry of Education (MOE 107QR001I5) in Taiwan.

Conflicts of Interest: The authors declare no conflicts of interest.

References

1. Monstrey, S.; Hoeksema, H.; Verbelen, J.; Pirayesh, A.; Blondeel, P. Assessment of burn depth and burn wound healing potential. *Burns* **2008**, *34*, 761–769. [CrossRef] [PubMed]
2. Heimbach, D.; Engrav, L.; Grube, B.; Marvin, J. Burn depth: A review. *World J. Surg.* **1992**, *16*, 10–15. [CrossRef]
3. Atiyeh, B.S.; Gunn, S.W.; Hayek, S.N. State of the art in burn treatment. *World J. Surg.* **2005**, *29*, 131–148. [CrossRef]
4. Pape, S.A.; Skouras, C.A.; Byrne, P.O. An audit of the use of laser Doppler imaging (LDI) in the assessment of burns of intermediate depth. *Burns* **2001**, *27*, 233–239. [CrossRef]
5. Droog, E.J.; Steenbergen, W.; Sjoberg, F. Measurement of depth of burns by laser Doppler perfusion imaging. *Burns* **2001**, *27*, 561–568. [CrossRef]
6. Holland, A.J.; Martin, H.C.; Cass, D.T. Laser Doppler imaging prediction of burn wound outcome in children. *Burns* **2002**, *28*, 11–17. [CrossRef]
7. Lawrence, C. Treating minor burns. *Nurs Times* **1989**, *85*, 69–73.
8. Gillitzer, R.; Goebeler, M. Chemokines in cutaneous wound healing. *J. Leukoc Biol.* **2001**, *69*, 513–521.
9. Pan, S.C.; Wu, L.W.; Chen, C.L.; Shieh, S.J.; Chiu, H.Y. Angiogenin expression in burn blister fluid: Implications for its role in burn wound neovascularization. *Wound Repair Regen.* **2012**, *20*, 731–739. [CrossRef]
10. Fett, J.W.; Strydom, D.J.; Lobb, R.R.; Alderman, E.M.; Bethune, J.L.; Riordan, J.F.; Vallee, B.L. Isolation and characterization of angiogenin, an angiogenic protein from human carcinoma cells. *Biochemistry* **1985**, *24*, 5480–5486. [CrossRef]
11. Katona, T.M.; Neubauer, B.L.; Iversen, P.W.; Zhang, S.; Baldridge, L.A.; Cheng, L. Elevated expression of angiogenin in prostate cancer and its precursors. *Clin. Cancer Res.* **2005**, *11*, 8358–8363. [CrossRef] [PubMed]
12. Pan, S.C.; Wu, L.W.; Chen, C.L.; Shieh, S.J.; Chiu, H.Y. Deep partial thickness burn blister fluid promotes neovascularization in the early stage of burn wound healing. *Wound Repair Regen.* **2010**, *18*, 311–318. [CrossRef] [PubMed]
13. Hawkes, R.; Niday, E.; Gordon, J. A dot-immunobinding assay for monoclonal and other antibodies. *Anal. Biochem.* **1982**, *119*, 142–147. [CrossRef]
14. Beyer, C.F. A 'dot-immunobinding assay' on nitrocellulose membrane for the determination of the immunoglobulin class of mouse monoclonal antibodies. *J. Immunol. Methods* **1984**, *67*, 79–87. [CrossRef]
15. Hawkes, R. The dot immunobinding assay. *Methods Enzymol.* **1986**, *121*, 484–491.
16. Cheng, C.M.; Martinez, A.W.; Gong, J.; Mace, C.R.; Phillips, S.T.; Carrilho, E.; Mirica, K.A.; Whitesides, G.M. Paper-based ELISA. *Angew. Chem. Int. Ed. Engl.* **2010**, *49*, 4771–4774. [CrossRef]
17. Hsu, M.Y.; Yang, C.Y.; Hsu, W.H.; Lin, K.H.; Wang, C.Y.; Shen, Y.C.; Chen, Y.C.; Chau, S.F.; Tsai, H.Y.; Cheng, C.M. Monitoring the VEGF level in aqueous humor of patients with ophthalmologically relevant diseases via ultrahigh sensitive paper-based ELISA. *Biomaterials* **2014**, *35*, 3729–3735. [CrossRef]
18. Wang, H.K.; Tsai, C.H.; Chen, K.H.; Tang, C.T.; Leou, J.S.; Li, P.C.; Tang, Y.L.; Hsieh, H.J.; Wu, H.C.; Cheng, C.M. Cellulose-based diagnostic devices for diagnosing serotype-2 dengue fever in human serum. *Adv. Healthc Mater* **2014**, *3*, 187–196. [CrossRef]
19. Cheng, J.Y.; Feng, M.J.; Wu, C.C.; Wang, J.; Chang, T.C.; Cheng, C.M. Development of a Sampling Collection Device with Diagnostic Procedures. *Anal. Chem.* **2016**, *88*, 7591–7596. [CrossRef]
20. Hsu, C.K.; Huang, H.Y.; Chen, W.R.; Nishie, W.; Ujiie, H.; Natsuga, K.; Fan, S.T.; Wang, H.K.; Lee, J.Y.; Tsai, W.L.; et al. Paper-based ELISA for the detection of autoimmune antibodies in body fluid-the case of bullous pemphigoid. *Anal. Chem.* **2014**, *86*, 4605–4610. [CrossRef]
21. Chan, W.C.W.; Udugama, B.; Kadhiresan, P.; Kim, J.; Mubareka, S.; Weiss, P.S.; Parak, W.J. Patients, Here Comes More Nanotechnology. *ACS Nano.* **2016**, *10*, 8139–8142. [CrossRef] [PubMed]
22. Marquez, S.; Morales-Narváez, E. Nanoplasmonics in Paper-Based Analytical Devices. *Front. Bioeng. Biotechnol.* **2019**, *7*, 69. [CrossRef] [PubMed]
23. Kishimoto, K.; Liu, S.; Tsuji, T.; Olson, K.A.; Hu, G.F. Endogenous angiogenin in endothelial cells is a general requirement for cell proliferation and angiogenesis. *Oncogene* **2005**, *24*, 445–456. [CrossRef] [PubMed]

24. Devgan, L.; Bhat, S.; Aylward, S.; Spence, R.J. Modalities for the assessment of burn wound depth. *J. Burns Wounds.* **2006**, *5*, e2.
25. Park, D.H.; Hwang, J.W.; Jang, K.S.; Han, D.G.; Ahn, K.Y.; Baik, B.S. Use of laser Doppler flowmetry for estimation of the depth of burns. *Plast Reconstr Surg.* **1998**, *101*, 1516–1523. [CrossRef]
26. O'Reilly, T.J.; Spence, R.J.; Taylor, R.M.; Scheulen, J.J. Laser Doppler flowmetry evaluation of burn wound depth. *J. Burn Care Rehabil.* **1989**, *10*, 1–6. [CrossRef]
27. Anselmo, V.J.; Zawacki, B.E. Multispectral photographic analysis. A new quantitative tool to assist in the early diagnosis of thermal burn depth. *Ann. Biomed. Eng.* **1977**, *5*, 179–193. [CrossRef]
28. Schneider, L.A.; Korber, A.; Grabbe, S.; Dissemond, J. Influence of pH on wound-healing: A new perspective for wound-therapy? *Arch. Dermatol Res.* **2007**, *298*, 413–420. [CrossRef]
29. Sharpe, J.R.; Booth, S.; Jubin, K.; Jordan, N.R.; Lawrence-Watt, D.J.; Dheansa, B.S. Progression of wound pH during the course of healing in burns. *J. Burn Care Res.* **2013**, *34*, e201–e208. [CrossRef]
30. Murphy, G.R.; Dunstan, R.H.; Macdonald, M.M.; Gottfries, J.; Roberts, T.K. Alterations in amino acid metabolism during growth by Staphylococcus aureus following exposure to H_2O_2—A multifactorial approach. *Heliyon* **2018**, *4*, e00620. [CrossRef]

Article

Requirements Analysis and Specification for a Molecular Tumor Board Platform Based on cBioPortal

Philipp Buechner [1], Marc Hinderer [1], Philipp Unberath [1], Patrick Metzger [2,3], Martin Boeker [4], Till Acker [5], Florian Haller [6], Elisabeth Mack [7], Daniel Nowak [8,9], Claudia Paret [10,11], Denny Schanze [12], Nikolas von Bubnoff [13,14,15], Sebastian Wagner [16], Hauke Busch [17], Melanie Boerries [2,18,19,*,†] and Jan Christoph [1,*,†]

1 Department of Medical Informatics, Friedrich-Alexander-University Erlangen-Nürnberg (FAU), 91058 Erlangen-Tennenlohe, Germany; philipp.buechner@fau.de (P.B.); marc.hinderer@fau.de (M.H.); philipp.unberath@fau.de (P.U.)
2 Institute of Medical Bioinformatics and Systems Medicine, Faculty of Medicine and Medical Center-University of Freiburg, 79110 Freiburg, Germany; patrick.metzger@mol-med.uni-freiburg.de
3 Faculty of Biology, University of Freiburg, 79104 Freiburg, Germany
4 Institute of Medical Biometry and Statistics, Faculty of Medicine and Medical Center, University of Freiburg, 79104 Freiburg, Germany; martin.boeker@imbi.uni-freiburg.de
5 Institute of Neuropathology, Justus-Liebig-University Giessen, 35392 Giessen, Germany; till.acker@patho.med.uni-giessen.de
6 Institute of Pathology, University Hospital Erlangen, 91054 Erlangen, Germany; florian.haller@uk-erlangen.de
7 Department of Hematology, Oncology and Immunology, Philipps-University Marburg, and University Hospital Gießen and Marburg, Marburg, Germany Baldingerstraße, 35043 Marburg, Germany; elisabeth.mack@staff.uni-marburg.de
8 Department of Hematology and Oncology, Medical Faculty Mannheim, Heidelberg University, 68167 Mannheim, Germany; daniel.nowak@medma.uni-heidelberg.de
9 Heinrich-Lanz-Center for Digital Health, Medical Faculty Mannheim, Heidelberg University, 68167 Mannheim, Germany
10 Pediatric Hematology/Oncology, Children's Hospital, University Medical Center of the Johannes Gutenberg-University Mainz, 55131 Mainz, Germany; paretc@uni-mainz.de
11 University Cancer Center (UCT) of the University Medical Center of the Johannes Gutenberg-University Mainz, 55131 Mainz, Germany
12 Institute of Human Genetics, University Hospital Magdeburg, Faculty of Medicine, Otto-von-Guericke University, 39120 Magdeburg, Germany; denny.schanze@med.ovgu.de
13 Department of Hematology and Oncology, Medical Center, University of Schleswig-Holstein, Campus Lübeck, 23538 Lübeck, Germany; nikolas.bubnoff@uniklinik-freiburg.de
14 German Cancer Consortium (DKTK), partner site Freiburg, 79106 Freiburg, Germany
15 Department of Hematology, Oncology and Stem Cell Transplantation, Medical Center, Faculty of Medicine, University of Freiburg, 79106 Freiburg, Germany
16 Department of Medicine 2, Hematology/Oncology, Goethe University Hospital, 60590 Frankfurt am Main, Germany; swagner@med.uni-frankfurt.de
17 Institute of Experimental Dermatology and Institute of Cardiogenetics, University of Lübeck, 23562 Lübeck, Germany; hauke.busch@uni-luebeck.de
18 Comprehensive Cancer Center Freiburg (CCCF), Faculty of Medicine and Medical Center – University of Freiburg, 79106 Freiburg, Germany
19 German Cancer Consortium (DKTK), partner site Freiburg; and German Cancer Research Center (DKFZ), 69120 Heidelberg, Germany
* Correspondence: melanie.boerries@uniklinik-freiburg.de (M.B.); jan.christoph@fau.de (J.C.)
† These authors contributed equally to this work.

Received: 15 December 2019; Accepted: 7 February 2020; Published: 10 February 2020

Abstract: Clinicians in molecular tumor boards (MTB) are confronted with a growing amount of genetic high-throughput sequencing data. Today, at German university hospitals, these data are usually handled in complex spreadsheets from which clinicians have to obtain the necessary information. The aim of this work was to gather a comprehensive list of requirements to be met by cBioPortal to support processes in MTBs according to clinical needs. Therefore, oncology experts at nine German university hospitals were surveyed in two rounds of interviews. To generate an interview guideline a scoping review was conducted. For visual support in the second round, screenshot mockups illustrating the requirements from the first round were created. Requirements that cBioPortal already meets were skipped during the second round. In the end, 24 requirements with sometimes several conceivable options were identified and 54 screenshot mockups were created. Some of the identified requirements have already been suggested to the community by other users or are currently being implemented in cBioPortal. This shows, that the results are in line with the needs expressed by various disciplines. According to our findings, cBioPortal has the potential to significantly improve the processes and analyses of an MTB after the implementation of the identified requirements.

Keywords: decision making, computer-assisted; decision support systems, clinical; precision medicine; computational biology; molecular tumor board; cBioPortal; requirements analysis; neoplasms

---

1. Introduction

Advances in next-generation sequencing (NGS) technology and the resulting decrease of costs hold large promises for personalized medicine, currently revolutionizing cancer diagnostics in particular. The sequencing of whole tumor exomes, genomes and transcriptomes of patients allows physicians to make individual molecular-guided decisions. However, the complex nature of cancer and its large interindividual heterogeneity require an interdisciplinary board composed of medical and scientific experts to review and interpret the equally complex analysis results. Recent data suggest that such molecular tumor boards (MTBs) have the potential to improve therapy and care for patients that have run out of guideline-based treatment options or have rare tumors [1,2].

Several German medical centers have already started to implement MTBs in their clinical environment, all working with various types of omics data from, e.g., NGS and other technologies [3,4]. To handle the many results from a large amount of omics data there is a high need for a standardized toolset that supports clinicians in analyzing and interpreting these data and creating high-quality presentations of complex multi-dimensional data effectively. Moreover, it requires both the integration of clinical with molecular and genomic data and the visualization of joint analysis results. However, experiences with the implementation and establishment of information technology (IT) and bioinformatics support for MTBs are still rare in Germany and probably world-wide, thus in need of improvement and optimization [5].

To address this issue, the MIRACUM consortium (Medical Informatics in Research and Care in University Medicine) has established the Use Case 3 which focuses on the provision of IT and bioinformatics support for translation and visualization of data analyzed in MTBs. As part of this use case, we will establish a generic, open-source framework that supports the analysis, interpretation and visualization of both clinical and omics data [6]. Data analysis is handled through MIRACUM-Pipe [7] which provides automatic, parameter-controlled processing of omics data with alignment, variant calling, annotation and analysis. The second aspect of the framework will be the data visualization and documentation of the results of the MTB. Both, the pipeline and the visualization, will be provided as separate and open-source components developed in a user-centered design process.

The cBio Cancer Genomics Portal (cBioPortal) was selected as a suitable platform to visualize the generated data supplied by the MIRACUM-Pipe [7] and to support the decision-making processes in an MTB. cBioPortal provides an extensive set of tools for exploring, visualizing and analyzing multi-dimensional and large-scale cancer genomics data sets [8,9]. Within the context of an MTB, cBioPortal may support case preparation, case review, and the documentation and communication of treatment recommendations in the near future. Therefore, it is well suited to replace the current practice at some German university hospitals of managing complex mutation data in huge spreadsheets by providing comprehensible visualizations [10,11]. At the participating clinics, cBioPortal could optimize the processing of up to 200 cases per year with sometimes hundreds of identified but not necessarily relevant mutations, and thus improve the decision making [5].

The integration of cBioPortal into the workflow of MTBs requires adjustments regarding different functionalities and needs (requirements). For instance, to meet some of our requirements the user needs to have write access to the data stored in cBioPortal. Therefore, we must find a proper solution to accomplish this in line with the concepts cBioPortal currently pursues as a (read-only) data warehouse.

The objective of this work is to provide a requirements specification for an IT platform based on cBioPortal that supports processes in molecular tumor boards to find and document a therapy recommendation. To our knowledge, there is so far no systematic assessment of such requirements from the point of view of MTB participants in different hospitals, even though there are already existing tools that integrate numerous data sources and even support the documentation process in a uniform MTB tool [12,13]. Our work could serve as a blueprint for the development of further tools based on cBioPortal for MTBs in Germany and worldwide.

2. Materials and Methods

To identify the requirements, we conducted a qualitative research to assess a set of potential requirements in two consecutive rounds of interviews. The second round of interviews became necessary because the first round was developed iteratively and therefore not all participants had the chance to comment on all mentioned potential requirements.

In preparation, we reviewed the literature published between 1997 and 2017 (scoped review) for existing systems, tools and knowledgebases that support molecular tumor boards (see Figure 1), resulting in an interview guideline for the first round of interviews.

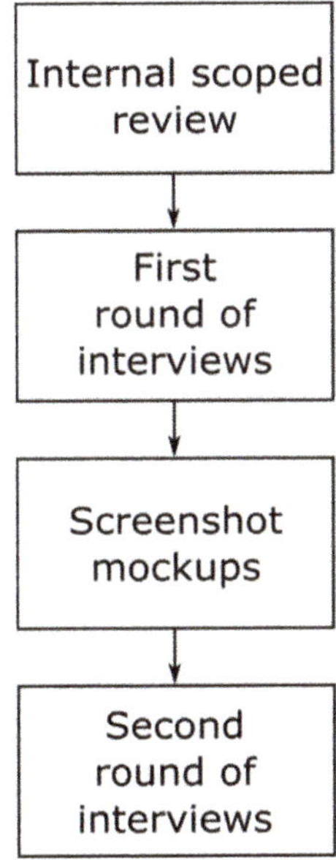

Figure 1. Outline of the process of requirements analysis.

Based on almost all assessments from the first round of interviews, for the second round, we created screenshot mockups for better understanding and visualization of possible implementations in cBioPortal.

2.1. Details about Scoping Review

We conducted a review of literature for potential MTB tools following the Preferred Reporting Items for Systematic Reviews and Meta-Analysis (PRISMA) guidelines [14,15] as far as appropriate for the requirements analysis. Therefore, we searched MEDLINE and Web of Science (all databases) for articles focusing on MTBs or equivalent clinical decision-making structures published between 1997 and 2017. We captured several features of different potential MTB tools which are either described in the literature or used by physicians at our MIRACUM sites. We used these findings to prepare for the first round of interviews. A detailed description of the methods we used can be found in Supplementary File 1.

2.2. First and Second Round of Interviews

Based on this prior knowledge, we conducted one group interview per partner site and per round from a constructivist point of view. We hypothesized that each site would have different views and visions on requirements for supporting a local MTB. Therefore, we took all suggestions regarding the demands of the participants into account.

The interviews were conducted in local focus groups, in which the interviewers served as moderators. This approach allowed discussions between the participants and thus as many requirements and their potential variants as possible could be identified. At each site, all participants were interviewed together in one session. The interviewers had an interdisciplinary background: medicine (Philipp Buechner, Melanie Boerries), medical informatics (Jan Christoph, Marc Hinderer), bioinformatics (Jan Christoph, Melanie Boerries) and biology (Melanie Boerries).

All interviews, except the one at the first author's local university hospital, took place as web conferences with transmission of voices and screen contents. In addition to the option of easily recording the session, the major argument for this setting was that the participants at the various university hospitals were spread all over Germany.

All interview participants were members of the MIRACUM Use Case 3 and thus known to us. The members responsible for the use case at each site arranged an appointment and invited additional local experts, who all were also members of the use case. All participants had to have experiences with the processes related to an MTB in order to join the interviews.

2.2.1. Structure and Purpose of the First Round of Interviews

Philipp Buechner, Marc Hinderer and Jan Christoph conducted the interviews of the first round together as members of MIRACUM's Use Case 3 between June 2018 and August 2018. It comprised a short guideline with questions (see Supplementary File 2) we developed from the results of the scoped review. We also demonstrated the main functionalities (see Supplementary Files 3 and 4) of the following potential MTB tools:

- cBioPortal [9],
- OncoKB [16],
- SOPHiA GENETICS [17],
- Clarivate "Key Pathway Advisor" [18],
- Clarivate "MetaCore" [18] and
- "CIViC" (Clinical Interpretations of Variants in Cancer) [19].

We collected all mentioned requirements cBioPortal must meet (including details about potential options for implementation) that were mentioned during the meetings. The interview process was developed iteratively and the information gained was immediately incorporated into the subsequent interviews with other partner sites during this first round. We used the web conference system "Adobe Connect" to conduct, record and subsequently analyze these interview sessions.

2.2.2. Structure and Purpose of the Second Round of Interviews

The second round of interviews was performed by Philipp Buechner, Melanie Boerries and Jan Christoph between November 2018 and December 2018. In order to make optimal use of the limited time during the interviews, Philipp Buechner developed a comprehensive interview guideline describing the requirements and their potential options identified in the first round with text and screenshot mockups (see Supplementary File 5). However, this round did not cover the requirements mentioned in the first round of interviews, that are already implemented in cBioPortal or are generally out of the scope of MIRACUM Use Case 3. To familiarize the participants with the requirements, this guideline—once it was finally validated by Melanie Boerries, Jan Christoph and Philipp Unberath —was handed out to them prior the meetings.

Since some requirements had more than one potential option regarding implementation and visualization, sites were asked to select one during this round of interviews. In case they had different opinions, they were encouraged to find a compromise.

For the final software specification—after all interviews have been conducted and analyzed - we grouped related features into larger meta-categories to account for individual requirements and yet to keep the assessment structured. For example, the term "sample metadata", comprises six (individual) data features:

- Localization and time of the sampling;
- Type of sampling (e.g., fine-needle aspiration biopsy);
- Distinction between fresh-frozen and formalin-fixed paraffin-embedded samples;
- Scope of sequencing (e.g., gene panel or whole-exome sequencing);
- Name and version of both the used panel and kit;
- Hyperlink to the corresponding product-specific website of the manufacturer.

When calculating the total number of requirements identified by us, we only counted those combined meta categories. Therefore, the above-mentioned example of "sample metadata" counts as one requirement instead of six individual ones.

We used the web conference system "Zoom" to conduct, record and subsequently analyze these interview sessions.

2.3. *Low-Fidelity Mockup Demonstrator*

We created 54 descriptive screenshot mockups for almost all options of the identified requirements from the first round of interviews using the image-editing tool GNU Image Manipulation Program (GIMP), version 2.10.8. These low-fidelity mockups are based on full-screen screenshots of the cBioPortal graphical user interface and have been manipulated to give the realistic appearance of providing the respective functions. To quickly direct the viewer's focus to the part of the image that represents the demanded function we indirectly highlighted the area by darkening the rest of the image with a black overlay (opacity: 20%).

2.4. *Consultation with Main Developers of MSKCC*

After all interviews have been conducted, we discussed the requirements with the main developers of cBioPortal from the Memorial Sloan Kettering Cancer Center (MSKCC) in New York, USA, in an online audio conference. Prior to that, we had detailed the most important and far-reaching changes in a letter including excerpts from our mockups.

The aim of this was to increase the chances of merging our planned implementations into the main development branch of cBioPortal and to maintain contact with the main developers right from the beginning.

2.5. Ethical Approval

This study was ethically approved by the ethics committee of the Friedrich-Alexander-University Erlangen-Nürnberg (FAU) (see Supplementary File 6).

3. Results

3.1. Overview of Scoped Review

Based on our keyword search we selected 306 unique articles out of which 27 dealt with MTBs and were kept for further review. Next, two papers were discarded because their full texts were not available. From the remaining 25 publications, we excluded another 13 since they did not describe IT support in MTBs, which resulted in a total of twelve articles for our review. For details see Supplementary File 1.

3.2. Details about Interviewees

We conducted the first round of interviews with a total of 18 participants at nine different university hospitals to determine all requirements for an MTB software tool based on cBioPortal. Up to four participants were interviewed simultaneously at each site. Representatives of the following disciplines were involved: oncology (10), pathology (4), systems medicine/systems biology (2), bioinformatics (1) and human genetics (1).

The second round of interviews discussed and evaluated the requirements identified in the first round in detail with a total of 16 participants at the same university hospitals as above. Up to three participants were interviewed simultaneously at each site. Representatives of the following disciplines were present: oncology (7), pathology (4), systems medicine/systems biology (2), bioinformatics (1), human genetics (1) and urology (1).

3.3. Requirements from the First Round of Interviews and Screenshot Mockups

During the first round of interviews, a total of 49 requirements with up to seven potential options for implementation each were surveyed (see Supplementary File 7). Ten requirements either already implemented in cBioPortal or out of scope of our Use Case were dropped prior to the second round of interviews. This included features to:

- Highlight mutations with existing treatment options;
- Display information about general availability of a specific drug in Germany;
- Point out mutations causing treatment resistance;
- Mark germline mutations;
- Display variant allele frequencies alongside corresponding coverage;
- Integrate the database "Clinical Interpretations of Variants in Cancer" (CIViC) [19];
- Visualize mRNA expression data.

A platform to discuss individual mutations across hospitals, for example in the context of a forum, is outside the scope of MIRACUM's Use Case 3 and was, therefore, not included in the second round of interviews either.

The choice of the file format to be used for the import of the mutation data into cBioPortal was made independently of the interviews by all use case members (MAF: "mutation annotation format"). In addition, a feature to permanently hide mutations in a specific sample in cBioPortal was denied since this filtering should be done by MIRACUM-Pipe [7] only.

In preparation for the second round of interviews, we created a total of 54 screenshot mockups demonstrating almost all surveyed requirements and their respective options (see Supplementary File 8).

3.4. Consolidated Requirements from the Second Round of Interviews

Below we provide a rough overview of the consolidated requirements we surveyed during the second round of interviews. For a list and detailed description of all final requirements and their respective options see Supplementary File 9.

3.4.1. Improving Patient Case Analysis

Since case analysis in personalized medicine relies on various information such as details about the patient or—in case of MTBs—the underlying tumor, cBioPortal should provide those by integrating various information and knowledge sources in a single tool. Therefore, the participants requested, amongst others, clinical patient data to be stored in cBioPortal. Displaying sample metadata and their subsequent analysis was also requested. This includes, for example, the type and location of a biopsy as well as the exact specifications of the sequencer used.

Furthermore, cBioPortal seems to lack important (calculated) values for its usage in molecular tumor boards. Ranging from the tumor mutational burden (see Figure 2A) up to values that indicate the pathogenicity of individual mutations (see Figure 2C). Mutations are automatically annotated with the latter by the MIRACUM-Pipe [7] and this information should be displayed in cBioPortal. We also discovered potential improvements for already existing features in cBioPortal, like the visualization of copy number variations (see Figure 2E).

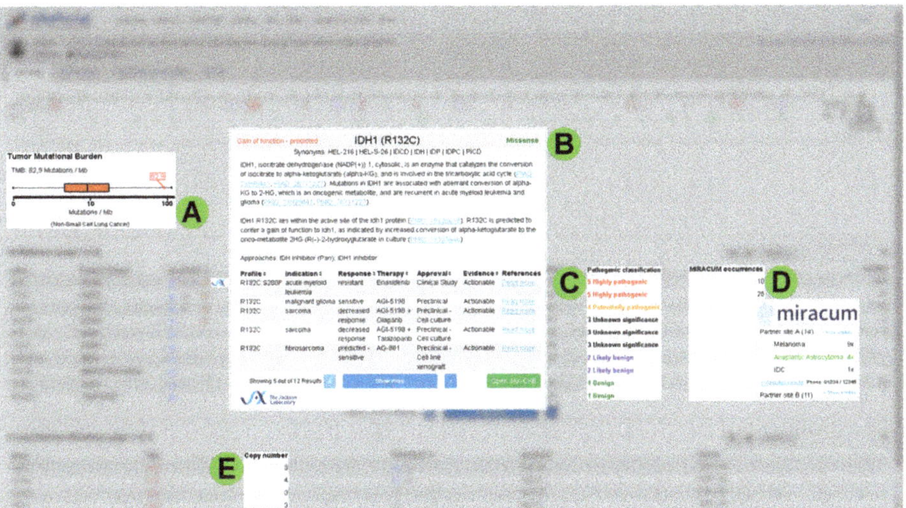

Figure 2. Collage demonstrating some requirements for the case analysis. This collage is a synthesis of different screenshot mockups, which exemplarily illustrates some requirements for individual case analysis: (**A**) A visual representation of the tumor mutational burden relative to the expected range of the corresponding tumor entity using a box plot. (**B**) Integration of data from the JAX Clinical Knowledgebase (JAX-CKB) via annotation and pop-up. (**C**) Classification of pathogenicity as an example of data extension in the mutation table of the patient view. (**D**) Tool for an entity-specific display of how often a certain mutation was found and classified as relevant for the therapy recommendation at the partner sites. If necessary, the contact data may be used to exchange experiences. (**E**) Extension of the table with copy number variation (CNV) data by the exact number of copies. Example data adopted from the public cBioPortal (https://cbioportal.org) and JAX-CKB knowledgebase [20].

Since the interviewees also use numerous different databases when evaluating a patient case, the integration of additional knowledgebases is highly desired. In this context, the JAX Clinical

Knowledgebase [20,21] was mentioned explicitly and considered to be beneficial when integrated (see Figure 2B).

Besides that, the visualization of molecular pathways can be an important tool that links individual mutations to molecular function and pathway in the search for a therapy option.

In order to improve cooperation, there were requests for a central service to (automatically) report a mutation that has led to a therapy recommendation before. If other participating hospitals similarly identify this mutation in one of their samples, it should be highlighted and contact details should be displayed (see Figure 2D) for detailed, personal exchange of expertise.

3.4.2. Supporting the Development and Recording of a Therapy Recommendation

Once the patient's data has been reviewed, the members of the molecular tumor board determine—if possible—the potentially relevant mutations for a therapy recommendation based on their previous analysis. This selection should be recorded in cBioPortal (see Figure 3A) and serve as the basis of the following features.

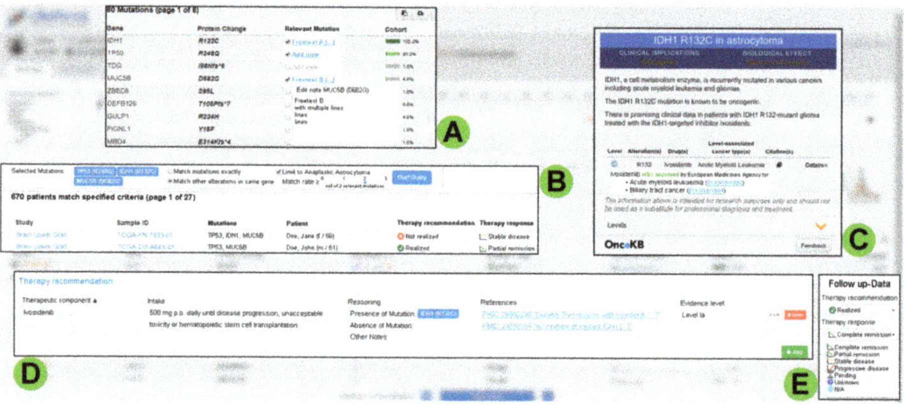

Figure 3. Collage demonstrating some requirements for the development and recording of a therapy recommendation. This collage depicts image sections from the screenshot mockups we created based on the original interface of cBioPortal: (**A**) Checkboxes and text fields in the mutation table of the patient view to mark potentially relevant mutations for the therapy recommendation. (**B**) Search functionality with automatic parameter transfer to find previous patient cases with a mutation pattern similar to that of the current patient. (**C**) Extension of OncoKB's information to cover the European Medicines Agency's (EMA) approval status of a given drug. (**D**) Summary of already entered components of the therapy recommendation for the current patient. (**E**) Option to record follow-up data for the current patient. Example data adopted from the public cBioPortal (https://cbioportal.org) and OncoKB [16].

This set of relevant mutations is used to search for similar patient cases that have been analyzed in the local hospital before. A search should be comprehensively parameterizable and values from the current patient case (e.g., tumor entity, relevant mutations, etc.) should be automatically applied (see Figure 3B). The interviewees considered the gathering of information related to therapy recommendations for previous patients as the main goal of this functionality. This requirement was complemented by the need to document follow-up data (see Figure 3E). Therefore, details on the progression status of similar patient cases can be reviewed and included in the evaluation of the current case more easily.

Building on this, further information on a possible therapy approach is required. In addition to the already available functions in cBioPortal provided by OncoKB [16], the interviewees requested a way to easily query the approval status of a drug in their respective country (in the case of MIRACUM: Germany) (see Figure 3C).

Since, according to the participants, some therapy components (e.g., a drug) are only available in the context of (pre-) clinical studies, integration of clinical trial databases such as ClinicalTrials.gov [22] were requested. The interviewees expect a more efficient search for suitable studies through the automatic transfer of relevant search parameters, which are taken from the cBioPortal data record.

Once a patient case has been prepared based on its individual data, it will be presented and discussed during a meeting of the MTB in order to jointly develop a therapy recommendation. There was no consensus as to what extent cBioPortal must support such a presentation. Some considered an automatically generated set of slides with individualized content as helpful. Others preferred to use the cBioPortal graphical user interface itself during the presentation with an option to hide irrelevant content. However, some interviewees also considered such assistance to be completely unnecessary.

After a therapy recommendation has been decided upon within the MTB, it must be recorded in detail in cBioPortal. Besides information on the therapy itself (e.g., the name of a drug), the molecular and clinical rationale for the recommendation needs to be documented, too. As a justification, the mutations earlier classified as relevant or the tumor mutational burden (TMB) may be referenced.

To submit the results of the MTB to the client (e.g., the treating physician), the interviewees requested a function to generate a PDF report. Besides the actual therapy recommendation, this report should also contain extracts from the consulted databases and thus also be used to archive the current state of knowledge that led to the decisions made.

3.4.3. Requirements for IT Infrastructure

In order to integrate cBioPortal in the various clinical system landscapes, a standardized application programming interface (API) like FHIR (Fast Healthcare Interoperability Resources [23]) should be used by the respective components of the hospital information system to feed a local cBioPortal instance with various and comprehensive data (e.g., clinical data). In addition, the information, which is created or altered within cBioPortal, should be accessible via this interface for export into the local hospital information system.

The existing user system in cBioPortal must be extended by a comprehensive and flexible user and rights management. Of course, only authenticated and authorized users should be permitted to use the cBioPortal instance for an MTB at all. Furthermore, it should be possible to "freeze" a therapy recommendation made by the molecular tumor board and to allow subsequent changes only for certain users and only in a reproducible way. This implies that all alterations to the data records in cBioPortal can be traced, thus guaranteeing their integrity and authenticity.

In addition, some sites indicated that patient data may be stored in a pseudonymized manner if this becomes necessary for legal reasons (like it is done in the public available cBioPortal instance hosted by the Memorial Sloan Kettering Cancer Center). Of course, in this case, it must be guaranteed that the data can be reassigned to the patient at the end. For this purpose, an identification code assigned to the patient by the hospital information system could be used for pseudonymization.

4. Discussion

Due to the huge amount of data produced by NGS technologies, the growing number of clinical studies and released targeted therapies to address treatment of cancer, new tools are required to identify relevant molecular alterations and matching therapy options for individual patients. The present paper outlines which of those are essential and which functionalities exactly have to be provided to support processes in MTBs. To our knowledge, there is no comparable work so far tackling this issue from the point of view of participants in MTBs.

4.1. Results and Future Work

We worked out a requirements specification for the software supporting the processes in an MTB based on the open source project cBioPortal, an integrated database, which allows to visualize clinical parameters and molecular findings of individual patients in the context of current knowledge about cancer diseases. It turned out that the screenshot mockups we had created prior to the second round of interviews played an important role in the further requirements analysis. They allowed us to describe the different requirements and their potential options identified in the first round of interviews briefly and succinctly and thus to quickly start a discussion in the second round. They may also serve as blueprints for the future implementation of the requirements as it was the case with the integration of the Genome Aggregation Database (gnomAD [24]), where our mockups served as inspiration for the core cBioPortal development [25].

This conversation with the participants turned out to be the most important part of this work at all, as on the one hand questions that occurred could be resolved immediately and on the other hand, requirements that had not yet been considered by neither the interviewees nor us came to light by intensive discussions. Therefore, we could finally identify 24 requirements that were not yet integrated into cBioPortal at the time of the survey. Some of them have been proposed to the community of cBioPortal with requests for comments (RFC) filed by other users or even have been implemented before the finalization of this paper [25–32]. RFCs are publicly available documents everyone can create with proposals for new features in cBioPortal that often also include descriptions of a potential way to implement these. They are an important step in the development of new functionalities in cBioPortal and help to involve the community in the process of implementation. This demonstrates, how important those requirements—even outside of an MTB context—seem to be for the future of cBioPortal.

Besides that, some of our requirements identified are also met by tools not directly associated with cBioPortal, like MatchMiner [33] published by the Dana–Farber Cancer Institute. This tool, which has recently been prototypically integrated in cBioPortal [34], offers a service to match clinical trials to a patient case. At the moment, all studies have to be cured manually, so to use this feature to prepare MTBs in Germany, the ongoing challenge of an automatic import from databases like ClinicalTrials.gov [22] still remains to be resolved.

For the future we plan to integrate cBioPortal in the hospital information systems (HIS) of our partner sites. This means the cBioPortal instance imports clinical data related to patients that have been registered for the MTB directly from the corresponding electronic health record (EHR). This data comprises at least age, gender and tumor entity and ideally also recent diagnoses and therapy attempts as well as data from cancer registries. The sequencing data is automatically processed by the MIRACUM-Pipe [7] under parameter control, including alignment, variant calling, annotation and analyses, in order to finally be passed directly to the cBioPortal instance.

In addition to the import of existing data into cBioPortal, users also need to add persistent information to patient cases in cBioPortal to mark relevant mutations in a patient's sample or to finally document the therapy recommendation. Since we are in ongoing contact with the main developers of cBioPortal, who finally determine which new features are merged into their project and provide future maintenance of them, we identified a considerable conflict with this requirement: cBioPortal's primary focus is to support research in form of a read-only data warehouse. Therefore, the main developers stressed, that having direct write access by the user is currently not intended. A solution for this problem would be to place a hyperlink in cBioPortal to an input mask not hosted in cBioPortal itself, where the user can enter the data (e.g., the therapy recommendation). This form forwards the data to the patient's EHR in the HIS via a standardized application programming interface (API) like FHIR [23], which is supported by most modern systems of an EHR (e.g., MEONA [35]). In order to make this data available again in cBioPortal, an (automatic) import must take place on a regular basis (e.g., twice daily). Thus, the data warehouse concept of cBioPortal would remain and no user write access in cBioPortal itself is necessary.

We also discussed another problem we came across: the rigid process to import mutation data into cBioPortal. At the moment, there is no way to (dynamically) import additional annotation data on individual mutations without fundamental changes to the backend, which would be necessary for the integration of the demanded scores and similar. The team in Use Case 3 of the MIRACUM consortium responsible for the implementation already took the first steps and submitted a request for comments to the community to address this problem. This request deals with a flexible integration of further mutation data by means of an additional database column with the data stored in JavaScript object notation (JSON) format [36].

In general, it is necessary to store data in a structured manner rather than in free text in order to achieve a high grade of automation during import. For example, the International Classification of Diseases for Oncology (ICD-O) can be used to encode the tumor entity, the German Federal Ministry of Health has stated in its announcement [37]. The use of a standardized ontology, such as that provided by OncoTree [38], can also be useful.

However, not only technical hurdles have to be tackled, but also legal ones. For example, when integrating further databases, particular attention must be paid to licensing as some restrict use to research purposes only. For example, even though the JAX Clinical Knowledgebase (JAX-CKB) was developed "to support clinical decision-making" [21], in the disclaimer of their website they allow usage "only for research and educational purposes" [20]. Use Case 3 of the MIRACUM consortium currently uses cBioPortal in clinical research. However, if it is used outside of a research context in the future, of course, this and also existing laws and regulations must be taken into account during the development, as well. Besides compliance with data protection regulations [39,40] and the Medical Devices Act [41], this is a very broad field.

As for future works, the implementation of the collected requirements must address these problems and find viable ways in close coordination with the main developers and the community around cBioPortal. This is the only way to integrate the features into the project permanently and to ensure their further maintenance and support.

4.2. Related Work

We came across two related works taking place in Germany. Halfmann et al. report that they developed a tool that aims to support the preparation process of a molecular tumor board as well as the presentation of a patient case during a meeting [12]. A video published by them demonstrates how different tools, including cBioPortal, can be combined in a single user interface [42]. Among other requirements, they discovered, like we did, that a function "to search for comparable local cases" [12] is demanded by clinical experts.

Fegeler et al. also describe a software solution for the support of molecular tumor boards. In addition to the option of planning and managing the processes in an MTB, they also describe an integrated video conferencing system. They plan to integrate cBioPortal and knowledge databases, to support the development of a therapy recommendation [13].

4.3. Limitations

Since we collected the requirements based on the expertise of clinicians who could only spend a certain amount of time for the survey and interviews (timeframe ranged between 40 min and two hours per interview and per round), we concentrated on the requirements which had not yet been implemented in cBioPortal. Therefore, we cannot make any statements about already existing features of the software that are useful for an MTB. As we iteratively improved and extended the interview guideline in the first round, the second round of interviews for consolidation had to be conducted so that in the end all sites had the opportunity to comment on every single requirement.

We also limited the number of sites interviewed to nine university hospitals spread all over Germany. Furthermore, not all disciplines and their representatives involved in a molecular tumor board were interviewed. Although we tried to incorporate as many disciplines as possible, this may not be a representative sample for Germany.

As a by-product of our requirements analysis and in preparation for the interviews, we performed a scoped review to provide an overview of already existing tools and systems that support molecular tumor boards. A limitation of it is its methodological rigor as compared to a full systematic review. We limited our database search to the MEDLINE and Web of Science databases. Therefore, we might have neglected relevant articles neither listed in MEDLINE nor in Web of Science. In addition, only one author performed the review, so this might also reduce the quality of results since misinterpretations cannot be systematically excluded. However, we believe that for our purposes we achieved a high degree of methodological quality throughout this scoping review by following the PRISMA statement [14,15] as far as appropriate for the requirements analysis.

5. Conclusions

By interviewing experts at our partner sites, we identified and consolidated for the first time a list of requirements for IT-supported preparation of molecular tumor boards based on cBioPortal. This list comprises a total of 24 requirements that had not yet been implemented during the time of the interviews. For almost all of them and their subordinated features, we have created descriptive screenshot mockups (54 in total) which supported the interview process and may contribute to the further development of cBioPortal. This work provides important information based on the clinical needs that will ultimately support the members of an MTB interpret the complex data for a personalized therapy recommendation.

Supplementary Materials: Supplementary materials can be found at http://www.mdpi.com/2075-4418/10/2/93/s1.

Author Contributions: Conceptualization, P.B., M.H., P.U., M.B. (Melanie Boerries) and J.C.; data curation, P.B. and M.H.; formal analysis, P.B. and M.H.; investigation, P.B., M.H., M.B. (Melanie Boerries) and J.C.; methodology, P.B., M.H. and J.C.; project administration, P.B., M.H. and J.C.; supervision, M.B. (Melanie Boerries) and J.C.; Validation, P.B., M.H., P.U., M.B. (Melanie Boerries) and J.C.; visualization, P.B.; writing—original draft, P.B.; writing—review and editing, P.B., M.H., P.U., P.M., M.B. (Martin Boeker), T.A., F.H., E.M., D.N., C.P., D.S., N.v.B., S.W., H.B., M.B. (Melanie Boerries) and J.C. All authors have read and agreed to the published version of the manuscript.

Funding: This research is supported by MIRACUM that is funded by the German Federal Ministry of Education and Research (BMBF) within the "Medical Informatics Funding Scheme" (FKZ 01ZZ1801A and FKZ 01ZZ1801B).

Acknowledgments: We especially thank all interview participants for their willingness and time to conduct the interviews. In addition, we would like to thank Nikolaus Schultz and Jianjiong Gao from the Memorial Sloan Kettering Cancer Center (MSKCC), who provided us with their feedback on the identified requirements and our planned changes to cBioPortal. The present work was performed in (partial) fulfillment of the requirements for obtaining the degree "Dr. med." from the Friedrich-Alexander-Universität Erlangen-Nürnberg (FAU) (Philipp Buechner).

Conflicts of Interest: The authors declare no conflicts of interest. The funders had no role in the design of the study; in the collection, analyses, or interpretation of data; in the writing of the manuscript, or in the decision to publish the results.

Abbreviations

API	Application programming interface
cBioPortal	cBio Cancer Genomics Portal
CIViC	Clinical Interpretation of Variants in Cancer (knowledgebase)
CNV	Copy Number Variation
EHR	Electronic Health Record
EMA	European Medicines Agency
FHIR	Fast Healthcare Interoperability Resources
GIMP	GNU Image Manipulation Program
gnomAD	Genome Aggregation Database
HIS	Hospital Information System
IT	Information technology
JAX-CKB	Jackson Laboratory Clinical Knowledgebase
JSON	JavaScript Object Notation
MIRACUM	Medical Informatics in Research and Care in University Medicine
MSKCC	Memorial Sloan Kettering Cancer Center

MTB	Molecular Tumor Board
NGS	Next-generation sequencing
PRISMA	Preferred Reporting Items for Systematic Reviews and Meta-Analysis
RFC	Request for Comments
TMB	Tumor Mutational Burden

References

1. Parker, B.A.; Schwaederlé, M.; Scur, M.D.; Boles, S.G.; Helsten, T.; Subramanian, R.; Schwab, R.B.; Kurzrock, R. Breast Cancer Experience of the Molecular Tumor Board at the University of California, San Diego Moores Cancer Center. *J. Oncol. Pract.* **2015**, *11*, 442–449. [CrossRef] [PubMed]
2. Bryce, A.H.; Egan, J.B.; Borad, M.J.; Stewart, A.K.; Nowakowski, G.S.; Chanan-Khan, A.; Patnaik, M.M.; Ansell, S.M.; Banck, M.S.; Robinson, S.I.; et al. Experience with precision genomics and tumor board, indicates frequent target identification, but barriers to delivery. *Oncotarget* **2017**, *8*, 27145–27154. [CrossRef] [PubMed]
3. Hoefflin, R.; Geißler, A.-L.; Fritsch, R.; Claus, R.; Wehrle, J.; Metzger, P.; Reiser, M.; Mehmed, L.; Fauth, L.; Heiland, D.H.; et al. Personalized Clinical Decision Making Through Implementation of a Molecular Tumor Board: A German Single-Center Experience. *JCO Precis. Oncol.* **2018**, 1–16. [CrossRef]
4. Perera-Bel, J.; Hutter, B.; Heining, C.; Bleckmann, A.; Fröhlich, M.; Fröhling, S.; Glimm, H.; Brors, B.; Beißbarth, T. From somatic variants towards precision oncology: Evidence-driven reporting of treatment options in molecular tumor boards. *Genome Med.* **2018**, *10*, 18:1–18:15. [CrossRef] [PubMed]
5. Hinderer, M.; Boerries, M.; Haller, F.; Wagner, S.; Sollfrank, S.; Acker, T.; Prokosch, H.-U.; Christoph, J. Supporting Molecular Tumor Boards in Molecular-Guided Decision-Making - The Current Status of Five German University Hospitals. *Stud. Health Technol. Inform.* **2017**, *236*, 48–54. [CrossRef] [PubMed]
6. Prokosch, H.-U.; Acker, T.; Bernarding, J.; Binder, H.; Boeker, M.; Boerries, M.; Daumke, P.; Ganslandt, T.; Hesser, J.; Höning, G.; et al. MIRACUM: Medical Informatics in Research and Care in University Medicine. *Methods Inf. Med.* **2018**, *57*, e82–e91. [CrossRef] [PubMed]
7. Metzger, P.; Scheible, R.; Hess, M.; Boeker, M.; Andrieux, G.; Börries, P. MIRACUM-Pipe (AG-Boerries/MIRACUM-Pipe Repository). Available online: https://github.com/AG-Boerries/MIRACUM-Pipe (accessed on 11 August 2019).
8. Gao, J.; Aksoy, B.A.; Dogrusoz, U.; Dresdner, G.; Gross, B.; Sumer, S.O.; Sun, Y.; Jacobsen, A.; Sinha, R.; Larsson, E.; et al. Integrative analysis of complex cancer genomics and clinical profiles using the cBioPortal. *Sci. Signal.* **2013**, *6*. [CrossRef] [PubMed]
9. Cerami, E.; Gao, J.; Dogrusoz, U.; Gross, B.E.; Sumer, S.O.; Aksoy, B.A.; Jacobsen, A.; Byrne, C.J.; Heuer, M.L.; Larsson, E.; et al. The cBio cancer genomics portal: an open platform for exploring multidimensional cancer genomics data. *Cancer Discov.* **2012**, *2*, 401–404. [CrossRef]
10. Shneiderman, B. Inventing Discovery Tools: Combining Information Visualization with Data Mining. In *Discovery science, Proceedings of the 4th International Conference, DS 2001, Washington, DC, USA, 25–28 November 2001*; Jantke, K.P., Ed.; Springer: Berlin/Heidelberg, Germany, 2001; pp. 17–28. ISBN 978-3-540-42956-2.
11. Omta, W.A.; de Nobel, J.; Klumperman, J.; Egan, D.A.; Spruit, M.R.; Brinkhuis, M.J.S. Improving Comprehension Efficiency of High Content Screening Data Through Interactive Visualizations. *Assay Drug Dev. Technol.* **2017**, *15*, 247–256. [CrossRef] [PubMed]
12. Halfmann, M.; Stenzhorn, H.; Gerjets, P.; Kohlbacher, O.; Oestermeier, U. User-Driven Development of a Novel Molecular Tumor Board Support Tool. In *Data Integration in the Life Sciences*; Auer, S., Vidal, M.-E., Eds.; Springer International Publishing: Cham, Switzerland, 2019; ISBN 978-3-030-06015-2.
13. Fegeler, C.; Zsebedits, D.; Bochum, S.; Finkeisen, D.; Martens, U.M. Implementierung eines IT-gestützten molekularen Tumorboards in der Regelversorgung. *FORUM* **2018**, *5*, 322–328. [CrossRef]
14. Moher, D.; Liberati, A.; Tetzlaff, J.; Altman, D.G. Preferred reporting items for systematic reviews and meta-analyses: the PRISMA statement. *PLoS Med.* **2009**, *6*, e1000097:1–e1000097:6. [CrossRef] [PubMed]

15. Liberati, A.; Altman, D.G.; Tetzlaff, J.; Mulrow, C.; Gøtzsche, P.C.; Ioannidis, J.P.A.; Clarke, M.; Devereaux, P.J.; Kleijnen, J.; Moher, D. The PRISMA statement for reporting systematic reviews and meta-analyses of studies that evaluate health care interventions: explanation and elaboration. *PLoS Med.* **2009**, *6*, e1000100:1–e1000100:28. [CrossRef] [PubMed]
16. Chakravarty, D.; Gao, J.; Phillips, S.M.; Kundra, R.; Zhang, H.; Wang, J.; Rudolph, J.E.; Yaeger, R.; Soumerai, T.; Nissan, M.H.; et al. OncoKB: A Precision Oncology Knowledge Base. *JCO Precis. Oncol.* **2017**. [CrossRef] [PubMed]
17. SOPHiA GENETICS. Available online: https://www.sophiagenetics.com/ (accessed on 29 May 2019).
18. Dubovenko, A.; Nikolsky, Y.; Rakhmatulin, E.; Nikolskaya, T. Functional Analysis of OMICs Data and Small Molecule Compounds in an Integrated "Knowledge-Based" Platform. *Methods Mol. Biol.* **2017**, *1613*, 101–124. [CrossRef] [PubMed]
19. Griffith, M.; Spies, N.C.; Krysiak, K.; McMichael, J.F.; Coffman, A.C.; Danos, A.M.; Ainscough, B.J.; Ramirez, C.A.; Rieke, D.T.; Kujan, L.; et al. CIViC is a community knowledgebase for expert crowdsourcing the clinical interpretation of variants in cancer. *Nat. Genet.* **2017**, *49*, 170–174. [CrossRef] [PubMed]
20. The Jackson Laboratory. JAX Clinical Knowledgebase - Disclaimer. Available online: https://ckb.jax.org/about/disclaimer (accessed on 18 August 2019).
21. Patterson, S.E.; Liu, R.; Statz, C.M.; Durkin, D.; Lakshminarayana, A.; Mockus, S.M. The clinical trial landscape in oncology and connectivity of somatic mutational profiles to targeted therapies. *Hum. Genomics* **2016**, *10*, 4:1–4:13. [CrossRef] [PubMed]
22. U.S. National Library of Medicine. ClinicalTrials.gov. Available online: https://clinicaltrials.gov/ (accessed on 10 August 2019).
23. HL7. Welcome to FHIR. Available online: https://www.hl7.org/fhir/ (accessed on 30 August 2019).
24. Karczewski, K.J.; Francioli, L.C.; Tiao, G.; Cummings, B.B.; Alföldi, J.; Wang, Q.; Collins, R.L.; Laricchia, K.M.; Ganna, A.; Birnbaum, D.P.; et al. Variation across 141,456 human exomes and genomes reveals the spectrum of loss-of-function intolerance across human protein-coding genes. *bioRxiv* **2019**. [CrossRef]
25. GitHub User Leexgh. Pull Request: Add Gnomad Column #2064. Available online: https://github.com/cBioPortal/cbioportal-frontend/pull/2064 (accessed on 15 August 2019).
26. GitHub User Leexgh. Pull Request: Add dbsnp Column to Mutation Table #2502. Available online: https://github.com/cBioPortal/cbioportal-frontend/pull/2502 (accessed on 15 August 2019).
27. GitHub User Leexgh. Pull Request: Add Clinvar to Patient Page #2596. Available online: https://github.com/cBioPortal/cbioportal-frontend/pull/2596 (accessed on 15 August 2019).
28. GitHub User Jjgao. Issue: Adding Clinical Trials Matching in Patient View #6444. Available online: https://github.com/cBioPortal/cbioportal/issues/6444 (accessed on 15 August 2019).
29. GitHub User Pvannierop. Pull Request: Integration of treatment response data in OncoPrint tab #2053. Available online: https://github.com/cBioPortal/cbioportal-frontend/pull/2053 (accessed on 15 August 2019).
30. GitHub User Pvannierop. Pull Request: Integration of Treatment Response Data in PlotsTab (Incl. New Waterfall Plot) #2055. Available online: https://github.com/cBioPortal/cbioportal-frontend/pull/2055 (accessed on 15 August 2019).
31. GitHub User Kjgao. Issue: PDF of Patient View Page #6446. Available online: https://github.com/cBioPortal/cbioportal/issues/6446 (accessed on 15 August 2019).
32. Lukasse, P.; van Hagen, S. RFC45: Gene Panel Information in Patient View. Available online: https://docs.google.com/document/d/1X7dA_wJFtv5xJO1oHCSt8DUdTmk07RexvHUpjCJsSM4/edit (accessed on 15 August 2019).
33. Lindsay, J.; Fitz, C.D.V.; Zwiesler, Z.; Kumari, P.; van der Veen, B.; Monrose, T.; Mazor, T.; Barry, S.; Albayrak, A.; Tung, M.; et al. MatchMiner: An Open Source Computational Platform for Real-Time Matching of Cancer Patients to Precision Medicine Clinical Trials Using Genomic and Clinical Criteria. Available online: https://www.biorxiv.org/content/10.1101/199489v3 (accessed on 20 January 2020).
34. GitHub User Victoria34. Pull Request: Matchminer Proxy Controller #5679. Available online: https://github.com/cBioPortal/cbioportal/pull/5679 (accessed on 21 January 2019).
35. MEONA GmbH. Meona. Available online: https://www.meona.de/ (accessed on 20 January 2020).
36. Unberath, P. RFC50: Add Support for Additional Arbitrary Variant Annotation. Available online: https://docs.google.com/document/d/1Pybk4_-lrirKJZ_cH64riZBRdWXdkJnCQqzx1O2fjRo/edit#heading=h.oyj1ec8k7lgx (accessed on 18 August 2019).

37. German Federal Ministry of Health. Updated Standardized Oncological Basic Data Set of the Consortium of German Tumor Centers e.V. (ADT) and the Society of Epidemiological Cancer Registries in Germany e.V. (GEKID) ("Aktualisierter Einheitlicher Onkologischer Basisdatensatz der Arbeitsgemeinschaft Deutscher Tumorzentren e.V. (ADT) und der Gesellschaft der Epidemiologischen Krebsregister in Deutschland e.V. (GEKID)"). Available online: https://www.bundesanzeiger.de/ (accessed on 9 March 2019).
38. OncoTree. Available online: https://github.com/cBioPortal/oncotree (accessed on 19 March 2019).
39. German Federal Ministry of Justice and Consumer Protection. German Federal Data Protection Act ("Bundesdatenschutzgesetz / BDSG"). Available online: https://www.gesetze-im-internet.de/englisch_bdsg/englisch_bdsg.pdf (accessed on 4 April 2019).
40. *REGULATION (EU) 2016/679 OF THE EUROPEAN PARLIAMENT AND OF THE COUNCIL - of 27 April 2016 - On the Protection of Natural Persons with Regard to the Processing of Personal Data and On the Free Movement of Such Data, and Repealing Directive 95/46/EC (General Data Protection Regulation)*; GDPR, 2016; European Parliament: Brussels, Belgium, 2016.
41. German Federal Ministry of Justice and Consumer Protection. German Medical Devices Act ("Gesetz über Medizinprodukte - MPG"). Available online: https://www.gesetze-im-internet.de/mpg/ (accessed on 18 August 2019).
42. YouTube-User PersOnS. Interface Prototype Demo—YouTube. Available online: https://www.youtube.com/watch?v=VXD3Rap11zg (accessed on 19 August 2019).

Article

Percussion Entropy Analysis of Synchronized ECG and PPG Signals as a Prognostic Indicator for Future Peripheral Neuropathy in Type 2 Diabetic Subjects

Hai-Cheng Wei [1,2], Na Ta [1], Wen-Rui Hu [1], Sheng-Ying Wang [1], Ming-Xia Xiao [1], Xiao-Jing Tang [3], Jian-Jung Chen [4,†] and Hsien-Tsai Wu [5,*,†]

1 School of Electrical and Information Engineering, North Minzu University, No. 204 North Wenchang Street, Yinchuan 750021, Ningxia, China; wei_hc@nun.edu.cn (H.-C.W.); ta_na@nmu.edu.cn (N.T.); 20187144@stu.nun.edu.cn (W.-R.H.); 20187150@stu.nun.edu.cn (S.-Y.W.); xiao_mx@nmu.edu.cn (M.-X.X.)

2 Basic Experimental Teaching and Engineering Training Center, North Minzu University, No. 204 North Wenchang Street, Yinchuan 750021, Ningxia, China

3 School of Science, Ningxia Medical University, No. 1160 Shengli Street, Yinchuan 750004, Ningxia, China; tangxj@nxmu.edu.cn

4 Taichung Tzuchi Hospital, The Buddhist Tzuchi Medical Foundation, Taichung 42743, Taiwan; cjjwei@tzuchi.com.tw

5 Department of Electrical Engineering, Dong Hwa University, No. 1, Sec. 2, Da Hsueh Rd., Shoufeng, Hualien 97401, Taiwan

* Correspondence: hsientsaiwu@gmail.com

† These authors contributed equally to this work.

Received: 11 November 2019; Accepted: 7 January 2020; Published: 9 January 2020

Abstract: Diabetic peripheral neuropathy (DPN) is one of the most common chronic complications of diabetes. It has become an essential public health crisis, especially for care in the home. Synchronized electrocardiogram (ECG) and photoplethysmography (PPG) signals were obtained from healthy non-diabetic ($n = 37$) and diabetic ($n = 85$) subjects without peripheral neuropathy, recruited from the diabetic outpatient clinic. The conventional parameters, including low-/high-frequency power ratio (LHR), small-scale multiscale entropy index (MEI_{SS}), large-scale multiscale entropy index (MEI_{LS}), electrocardiogram-based pulse wave velocity (PWV_{mean}), and percussion entropy index (PEI), were computed as baseline and were then followed for six years after the initial PEI measurement. Three new diabetic subgroups with different PEI values were identified for the goodness-of-fit test and Cox proportional Hazards model for relative risks analysis. Finally, Cox regression analysis showed that the PEI value was significantly and independently associated with the risk of developing DPN after adjustment for some traditional risk factors for diabetes (relative risks = 4.77, 95% confidence interval = 1.87 to 6.31, $p = 0.015$). These findings suggest that the PEI is an important risk parameter for new-onset DPN as a result of a chronic complication of diabetes and, thus, a smaller PEI value can provide valid information that may help identify type 2 diabetic patients at a greater risk of future DPN.

Keywords: type 2 diabetes; diabetic peripheral neuropathy (DPN); electrocardiogram (ECG); photoplethysmography (PPG); percussion entropy index (PEI)

1. Introduction

The prevention and early detection of diabetes mellitus (DM) in high-risk subjects such as those with metabolic syndrome have become important issues in preventive medicine and public health [1,2]. As for type 2 diabetic patients, the outcomes of microvascular and macrovascular complications can be grievous. Diabetic peripheral neuropathy (DPN) is not only one of the most common complications of chronic diabetes, but also a leading cause for disability due to foot amputation and ulceration, fall-related injury, and gait disturbance. [3,4]. Not only is the quality of life much lower among DPN patients but the mortality rate has also been shown to be higher than for DM alone [5]. To avoid developing a severe condition, blood glucose control is the most important issue for type 2 diabetic patients [6–8]. However, this is not easy to achieve, which is the reason many researchers have tried to develop some non-invasive methods and instruments [9] for type 2 diabetic patients to avoid the increased risk of developing atherosclerosis and autonomic nervous dysfunction [10–13].

Heart rate variability (HRV) using electrocardiography (ECG) is an assessment method of autonomic function and baroreflex sensitivity (BRS) [14]. The low-/high-frequency power ratio (LHR) is the autonomic function index for frequency domain analysis and it is considered to balance the reflection between sympathetic and parasympathetic activity changes [14–16]. However, the conventionally used frequency domain parameter is not always adequate for this purpose because of non-stationarity and the nonlinear physiological signals adopted [17–19]. Therefore, several new parameters based on nonlinear dynamics theory were recently applied to HRV studies on autonomic function and BRS [20–22]. Among these parameters, a small-scale multiscale entropy index (MEI_{SS}) was adopted to evaluate the autonomic nervous activities and BRS based on a nonlinear approach, using only the R–R interval (RRI) datasets [20]. In recent years, a new percussion entropy index (PEI) using synchronized ECG and photoplethysmography (PPG) signals was used to assess the BRS complexity in healthy elderly and diabetic subjects related to autonomic function changes [21,22]. Moreover, in a recent study [23] on a modified PEI, the PPG signals were measured and the peak-to-peak interval (PPI) series were calculated at the fingertip as an indicator of DPN in the aged and diabetic patients. However, real-time processing was not possible for the PPI-based index presented in [23]. On the other hand, parameters including the large-scale multiscale entropy index (MEI_{LS}) [20], in addition to pulse wave velocity (PWV_{mean}) [24–27], were reported for atherosclerosis detection in type 2 diabetes mellitus.

From the viewpoint of data analysis, we propose that some of the above parameters (i.e., LHR, MEI_{SS}, MEI_{LS}, PWV_{mean}, and PEI) could present highly significant differences between diabetic patients who underwent a synchronized ECG and PPG signals testing at baseline during the early stages of disease; these patients were then followed for a further six years, with and without DPN. Furthermore, this study was designed to apply synchronized ECG and PPG signals from a non-invasive instrument in predicting the development of peripheral neuropathy in type 2 diabetic patients. It is worth mentioning that PEI [21,22] has recently been introduced to assess the complexity of BRS, while the significance of smaller PEI values concerning the identification of subjects with type 2 diabetes who are more prone to developing diabetic neuropathy is unknown. This study aimed at testing the hypothesis that a smaller PEI value at the baseline measurement can provide valid information that may help identify type 2 diabetic patients at a greater risk of future DPN.

The rest of this paper is organized as follows. Section 2 describes the study population and the baseline examinations and protocol of measuring of synchronized ECG and PPG signals. A follow-up procedure and DPN status checking were addressed, followed by an explanation of the statistical analysis methods. As presented in Section 3, results from the five indices—LHR, MEI_{SS}, MEI_{LS}, PWV_{mean}, and PEI—were first computed for diabetic subjects with peripheral neuropathy within six years (i.e., Group 3) for comparison with the healthy elderly subjects (i.e., Group 1) and diabetic patients without peripheral neuropathy (i.e., Group 2). Subsequently, three new diabetic subgroups (i.e., Groups A–C) with different periods of PEI values were identified for follow-up and Cox proportional Hazards model as well as goodness-of-fit test for relative risks analysis. Finally, a Cox regression analysis of risk factors for the incidence of DPN within six years of follow-up in diabetic patients was verified. In Sections 4 and 5, the discussion of the results and conclusions from the present study are summarized, with suggestions for future work.

2. Materials and Methods

2.1. Study Design and Study Population

2.1.1. The Inclusion and Exclusion Criteria Were as Follows

Between July 2010 to March 2013, 128 subjects were enrolled for this study. All diabetic patients were recruited from the diabetes outpatient clinic of the Hualien Hospital (Hualien City, Taiwan), while healthy controls were recruited from a physical check-up program at the same hospital. All of the age-controlled healthy subjects had no personal or family history of cardiovascular diseases. Of the 128 volunteers, 6 were excluded due to a history of coronary heart disease, heart failure, ischemic stroke, peripheral arterial disease, chronic atrial fibrillation, or permanent pacemaker implantation.

2.1.2. Grouping

The remaining 122 subjects were then first divided into three groups, namely, healthy upper-middle-aged subjects (Group 1, age range: 40–79, number = 37), upper-middle-aged subjects diagnosed as having type 2 diabetes (Group 2, age range: 37–82, number = 58, glycated hemoglobin (HbA1c) ≥ 6.5%), and type 2 diabetic patients developing peripheral neuropathy within 6 years after baseline measurement (Group 3, age range: 44–77, number = 27) (Table 1). The baseline characteristics of these study subjects are presented in Table 1. Type 2 diabetes was diagnosed by either a fasting blood glucose concentration ≥126 mg/dL or HbA1c ≥ 6.5% [28]. Subsequently, the PEI values for diabetic patients—85 in total—in Groups 2 and 3 were arbitrarily divided into three new subgroups for the prognostication of subjects with type 2 diabetes who are more prone to develop DPN. That is, the diabetic subgroups were created on the basis of quartiles in the diabetic population distribution of the PEI—the upper 25% (i.e., Group A), the middle 50% (i.e., Group B), and the lower 25% (i.e., Group C).

Table 1. Anthropometric, hemodynamic, and serum biochemical parameters of the testing subjects.

Parameters	Group 1 (n = 37) Female/Man (19/18)	Group 2 (n = 58) Female/Man (24/34)		Group 3 (n = 27) Female/Man (13/14)	
Age, years	59.20 ± 1.67	61.80 ± 1.45	(p = 0.66)	62.81 ± 1.71	(p = 0.22)
Body height, cm	161.10 ± 1.19	160.37 ± 1.04	(p = 0.65)	164.15 ± 1.78	(p = 0.06)
Body weight, kg	60.95 ± 1.66	68.86 ± 1.45 **	(p = 0.00)	72.48 ± 1.46	(p = 0.13)
WC, cm	82.19 ± 1.77	93.32 ± 1.30 **	(p = 0.00)	96.50 ± 1.45	(p = 0.14)
BMI, kg/m^2	23.47 ± 0.59	26.82 ± 0.57 **	(p = 0.00)	27.08 ± 0.75	(p = 0.79)
SBP, mmHg	123.17 ± 3.21	125.96 ± 2.28	(p = 0.47)	127.85 ± 6.27	(p = 0.73)
DBP, mmHg	74.69 ± 1.55	74.91 ± 1.29	(p = 0.91)	73.23 ± 3.45	(p = 0.58)
PP, mmHg	48.49 ± 2.41	50.13 ± 2.08	(p = 0.61)	54.62 ± 3.73	(p = 0.26)
HDL, mg/dL	52.01 ± 3.59	44.65 ± 2.62	(p = 0.09)	42.79 ± 3.77	(p = 0.69)
LDL, mg/dL	114.65 ± 5.08	120.77 ± 6.48	(p = 0.49)	106.58 ± 4.90	(p = 0.15)
Cholesterol, mg/dL	192.16 ± 8.33	177.02 ± 7.65	(p = 0.19)	183.60 ± 7.01	(p = 0.58)
Triglyceride, mg/dL	92.45 ± 6.08	144.61 ± 11.44 **	(p = 0.00)	161.04 ± 13.47	(p = 0.39)
HbA1c, %	5.90 ± 0.06	8.12 ± 0.23 **	(p = 0.00)	8.36 ± 0.30	(p = 0.54)
FPG, mg/dL	99.80 ± 4.42	149.46 ± 6.59 **	(p = 0.00)	161.44 ± 11.26	(p = 0.33)

Values are expressed as mean ± SD. Group 1, healthy elderly subjects; Group 2, diabetic subjects; Group 3, diabetic subjects with peripheral neuropathy 6 years after baseline measurement. The total number of test subjects was 122. WC, waist circumference; BMI, body mass index; SBP, systolic blood pressure; DBP, diastolic blood pressure; PP, pulse pressure; HDL, high-density lipoprotein cholesterol; LDL, low-density lipoprotein cholesterol; HbA1c, glycosylated hemoglobin; FPG, fasting plasma glucose. ** $p < 0.001$ Group 1 vs. Group 2. p values of the parameter larger than 0.017 are regarded as not statistically significant between two groups. The total number of subjects is 122 in this table.

2.1.3. Ethical Issues, IRB, and Consent Form

The study was approved by the Institutional Review Board (IRB) of Hualien Hospital (Hualien City, Taiwan) [26,27] and Ningxia Medical University (Yinchuan City, Ningxia Province, China) Hospitals (No.2018-229). All subjects gave written informed consent.

2.1.4. Study Protocol

Each subject was required to refrain from caffeine-containing beverages and theophylline-containing medications for at least 8 h before the baseline data measurement. Before taking the measurements, all subjects were requested to sign informed consent forms and complete questionnaires on demographics and medical histories, as well as receive blood sampling for serum biochemical analysis.

A detailed explanation of the aim and procedures was provided, as well as the measurement of synchronized ECG and PPG signals to be used for follow-up study. The test subjects received a standardized medical examination by a doctor that consisted of anthropometric, physiological, and biochemical measures at baseline. The indices of atherosclerosis and autonomic nervous function were subsequently computed. All diabetic patients underwent regular clinic treatment and follow-up in an outpatient clinic for at least eight years (i.e., two years for DM identification and six years for the follow-up period). DPN was diagnosed as the presence of symptoms of numbness, tingling, or pain of distal extremities lasting for more than 3 months in the same diabetes outpatient department through neurophysiological study [29].

2.1.5. Follow-up and DPN Status

The DPN status for the subjects in Group 3 at each follow-up stage was ascertained by questionnaire and clinical medical examinations. The screening DPN from type 2 diabetes patients at the baseline and follow-up periods was based on the presence of symptoms of numbness, tingling, or pain of distal extremities lasting for more than 3 months and a confirmed diagnosis by the clinic doctor (i.e., in accordance with neurophysiological study). The study population comprised a sample of 27 type 2 diabetes patients with DPN (aged 62.81 ± 1.71 years) who underwent a synchronized ECG and PPG signals measurement at baseline and then were followed for at least 6 years after the baseline measurement at the same hospital.

2.2. Baseline Measurements and Protocol of Measurement of Synchronized Electrocardiogram (ECG) and Photoplethysmography (PPG) Signals

All measurements were performed over a period in the morning (i.e., 08:30–10:30). In addition, to minimize the potential errors in the infrared sensor readings arising from involuntary vibrations of the participants, all subjects were allowed to rest in a supine position for 30 min in a quiet room with a temperature maintained at 26 ± 1 °C. Blood pressure readings were obtained once over the left arm of the supine subjects using an automated oscillometric device (BP 3AG1; Microlife Corporation, Taipei, Taiwan) with an appropriate cuff size. A self-developed six-channel electrocardiography ECG-PWV-based system [26,27] was used to acquire 1000 successive recordings of the RRI signals and digital volume pulses (DVPs) within 30 min. To validate the application of the ECG-PWV system in assessing autonomic function, the RRI series was used for the LHR [14,16], MEI_{LS}, and MEI_{SS} computations. Accordingly, the present study analyzed the RRI signals by dividing the MEI according to a small scale (MEI_{SS}, mean value of sample entropy on a scale from 1 to 5) and large scale (MEI_{LS}, mean value of sample entropy on a scale from 6 to 10) for comparison [20]. The DVPs of PPG with the R wave on RRI as a reference point could be used for the electrocardiogram-based pulse wave velocity (PWV_{mean}) [26,27] and PEI [21,22] computations for assessing the autonomic function considering the degree of atherosclerotic change and autonomic function, respectively (Figure 1). In our previous study [21], changes in the BRS caused one to five cardiac cycle delays under the effects of fingertip DVP amplitude variations followed by synchronized RRIs. Accordingly, in obedience with the fluctuation tendency in the PEI computation, the percussion entropy had a length of the fluctuation pattern equal to two (i.e., PEI main contributor), and was expressed as BRS, while the percussion entropy, with a length of the fluctuation pattern equal to three (i.e., PEI major offset), was indicated as the biological complex system.

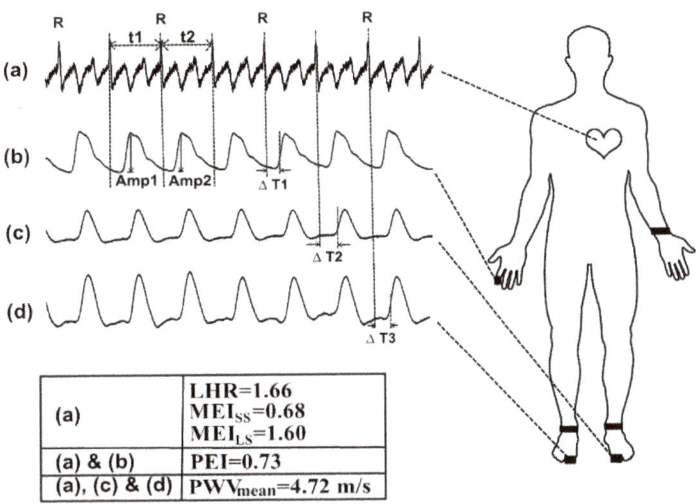

(a)	LHR=1.66 MEI_{SS}=0.68 MEI_{LS}=1.60
(a) & (b)	PEI=0.73
(a), (c) & (d)	PWV_{mean}=4.72 m/s

Figure 1. Schematic illustration of the measurements of six-channel electrocardiogram-based pulse wave velocity (ECG-PWV). The ECG and digital volume pulses (DVP) signals from one representative female subject in Group 1 with age of 44 and HbA1c of 5.1%. With (**a**) the R wave on Lead II of ECG, three parameters (i.e., low-/high- frequency power ratio (LHR), small-scale multiscale entropy index (MEI_{SS}), and large-scale multiscale entropy index (MEI_{LS}) were computed using only the RRI dataset. The synchronized ECGs (**a**) and the right index finger photoplethysmography (PPG) signals (**b**) were obtained for percussion entropy index (PEI) computation. With (**a**) the R wave on Lead II as a reference point, the time differences (ΔT_2 (**c**) and ΔT_3 (**d**)) for the second toe were obtained. The PWV_{mean} was calculated by dividing the distances from different points of reference (L) with ΔT (i.e., PWV = L/ΔT). The PWV_{mean}, in the evaluation of the degree of atherosclerosis in the lower extremity of the body, was obtained by averaging the PWV values from both sides of the foot.

2.3. Statistical Analysis

The values are expressed as mean ± SD in Tables 1–3. The comparisons of the continuous valuables were analyzed using a Student's unpaired *t* test with Bonferroni correction, and the differences between the categorical variables were assessed using a chi-square test. For the goodness-of-fit test and relative risk analysis, the PEI values were arbitrarily divided into three categories by the interquartile range method. The PEI was processed as both continuous and categorical variables and was undertaken in the Cox proportional hazards model to analyze the multivariate parameters according to the expansion of DPN. The relative risks (RR) were predicted with Cox regression analysis with corresponding 95% confidence intervals [30]. The following traditional risk factors for DPN were included as variables in the model: age, BMI, resting systolic and diastolic blood pressure, total cholesterol, triglyceride, waist circumference, pulse pressure, high-density lipoprotein cholesterol, low-density lipoprotein cholesterol, glycosylated hemoglobin, and fasting plasma glucose from baseline (i.e., PEI provided) to the end of the follow-up period (no longer than 6 years for each patient). According to the chi-square goodness-of-fit test in SPSS, the null hypothesis was rejected with a computed chi-square value larger than the level of significance. The Statistical Package for the Social Sciences (SPSS, version 14.0 for Windows, SPSS Inc., Chicago, IL, USA) was utilized for all statistical analyses.

3. Results

The results from the fours indices, LHR, MEI_{SS}, MEI_{LS}, PWV_{mean}, and PEI, were first computed for the diabetic subjects with peripheral neuropathy within six years (i.e., Group 3) for comparison with the healthy elderly subjects (i.e., Group 1) and diabetic patients without peripheral neuropathy (i.e., Group 2). Subsequently, three new diabetic subgroups using different PEI values were identified for the goodness-of-fit test. Finally, the Cox regression analysis of risk factors for the incidence of DPN within six years after the PEI provided for diabetic patients was verified.

3.1. Comparison among LHR, MEI_{SS}, MEI_{LS}, PWV_{mean}, and PEI for Age-Controlled Healthy and Diabetic Subjects with and without DPN

After the entire follow-up process had been carried out and the DPN status was confirmed, the results from the comparison of the four previous computational parameters (i.e., LHR, MEI_{SS}, MEI_{LS}, and PWV_{mean}) with the PEI for DPN identification assessment among the three groups of subjects are shown in Table 2. Although the value of PWV_{mean} was significantly higher in Group 2 compared with the Group 1 subjects ($p < 0.017$), there was no notable difference between Groups 2 and 3. On the other hand, the PEI showed highly significant differences among the three groups ($p < 0.001$) (Table 2).

Table 2. Of computational parameters for autonomic function assessment in three groups of testing subjects.

Parameters	Group 1 (n = 37)	Group 2 (n = 58)		Group 3 (n = 27)	
LHR	1.56 ± 0.17	2.00 ± 0.26	(p = 0.23)	2.34 ± 0.44	(p = 0.49)
MEI_{ss}	0.62 ± 0.08	0.57 ± 0.02 *	(p = 0.01)	0.55 ± 0.16	(p = 0.72)
MEI_{Ls}	1.56 ± 0.06	1.48 ± 0.04	(p = 0.31)	1.37 ± 0.06	(p = 0.69)
PWV_{mean}	4.65 ± 0.06	4.93 ± 0.06 *	(p = 0.01)	4.80 ± 0.07	(p = 0.22)
PEI	0.73 ± 0.01	0.63 ± 0.01 **	(p = 0.00)	0.59 ± 0.02 †	(p = 0.01)

Values are expressed as mean ± SD. Group 1, healthy elderly subjects; Group 2, diabetic subjects; Group 3, diabetic subjects with peripheral neuropathy within six years after baseline measurement. LHR, low- to high-frequency power ratio; MEI_{SS}, small-scale multiscale entropy index, MEI_{LS}, large-scale multiscale entropy index, PWV_{mean}, ECG-PWV-based pulse wave velocity; PEI, percussion entropy index. * $p < 0.017$ (p corrected), Group 1 vs. Group 2; ** $p < 0.001$, Group 1 vs. Group 2; † $p < 0.017$, Group 2 vs. Group 3. The total number of subjects in this table is 122.

3.2. Three Diabetic Subgroups Using Different Percussion Entropy Index (PEI) Values

The distribution of the PEI exhibited an approximately normal curve with a mild skew toward higher values. The values of the quartile ranges in the distribution were 0.27–0.54, 0.55–0.65, and 0.66–0.82 for the lower, middle two, and upper quartiles, respectively, for the prognostication of subjects with type 2 diabetes who are more prone to develop DPN. The diabetic patients in Group C showed remarkably higher HbA1c levels than those in the diabetic patients in Group B ($p < 0.017$). On the other hand, no significant differences were noted in the demographic and hemodynamic parameters, as well as the fasting blood glucose and serum lipid profile between any two groups (Table 3). In summary, a comparison of characteristics among subjects in the three categories revealed no significant differences in age, body mass index, waist circumference, systolic blood pressure, diastolic blood pressure, pulse pressure, high-density lipoprotein cholesterol, low-density lipoprotein cholesterol, and fasting plasma glucose.

3.3. Goodness-of-Fit Test and Cox Proportional Hazards Model for Relative Risks Analysis

3.3.1. The Goodness-of-Fit Test

According to the chi-square goodness-of-fit test result in SPSS, the null hypothesis (i.e., no association between the PEI and DPN) was rejected with a chi-square value (i.e., the computed chi-square value, $\chi^2 = 8.00$) larger than the level of significance ($\chi^2 = 5.99$, $\alpha = 0.05$). That is, diabetic patients with a smaller PEI value were associated with the future development of DPN within six years after the PEI was provided.

Table 3. Demographic, anthropometric, hemodynamic, and serum biochemical parameters of the testing diabetic patients in Groups 2 and 3.

Parameters	Group A (n = 22) Female/Man (8/14)	Group B (n = 42) Female/Man (20/22)		Group C (n = 21) Female/Man (9/12)	
Age, year	65.62 ± 1.62	64.95 ± 2.22	(p = 0.36)	63.00 ± 2.30	(p = 0.81)
Body height, cm	160.99 ± 1.32	162.90 ± 1.86	(p = 0.82)	161.53 ± 1.87	(p = 0.41)
Body weight, kg	68.98 ± 1.58	70.53 ± 9.37	(p = 0.32)	71.79 ± 2.30	(p = 0.57)
WC, cm	93.56 ± 1.51	94.10 ± 1.78	(p = 0.29)	96.32 ± 2.00	(p = 0.83)
BMI, kg/m^2	26.67 ± 0.61	26.61 ± 0.78	(p = 0.38)	27.73 ± 1.12	(p = 0.95)
SBP, mmHg	125.38 ± 4.13	130.63 ± 4.83	(p = 0.97)	125.16 ± 2.84	(p = 0.45)
DBP, mmHg	72.71 ± 2.26	75.58 ± 2.66	(p = 0.26)	76.79 ± 1.79	(p = 0.46)
PP, mmHg	52.67 ± 2.72	52.30 ± 4.53	(p = 0.32)	48.37 ± 2.10	(p = 0.94)
HDL, mg/dL	45.11 ± 3.19	46.67 ± 5.04	(p = 0.22)	38.88 ± 2.29	(p = 0.79)
LDL, mg/dL	110.30 ± 6.04	122.11 ± 11.42	(p = 0.27)	122.12 ± 8.64	(p = 0.32)
Cholesterol, mg/dL	171.53 ± 7.29	186.47 ± 10.13	(p = 0.22)	188.83 ± 13.38	(p = 0.25)
Triglyceride, mg/dL	148.97 ± 10.80	139.89 ± 14.50	(p = 0.57)	162.28 ± 24.77	(p = 0.63)
HbA1c, %	7.83 ± 0.24	8.40 ± 0.41	(p = 0.03)	8.78 ± 0.32 †	(p = 0.01)
FPG, mg/dL	148.84 ± 8.27	158.56 ± 13.71	(p = 0.47)	158.82 ± 8.78	(p = 0.53)

Group A, diabetic subjects with large PEI (the upper 25%); Group B, diabetic subjects with moderate PEI (the middle 50%); Group C, diabetic subjects with small PEI (the lower 25%). Values are expressed as mean ± SD. WC, waist circumference; BMI, body mass index; SBP, systolic blood pressure; DBP, diastolic blood pressure; PP, pulse pressure; HDL, high-density lipoprotein cholesterol; LDL, low-density lipoprotein cholesterol; HbA1c, glycosylated hemoglobin; FPG, fasting plasma glucose. † $p < 0.017$ (p corrected), Group B vs. Group C. The total number of patients in this table is 85.

3.3.2. Cox Proportional Hazards Model

A total of 27 type 2 diabetic patients developed DPN among 85 study patients (31.8%) within six years after the baseline examinations in this study. The progression to DPN in patients in the three categories within six years and corresponding relative risks for the incidence of DPN assessed by the Cox proportional hazards model are shown in Table 4. The Cox model revealed a graded association, with the diabetic subjects with a small PEI (i.e., Group C) at 2.90× greater risk of developing DPN on follow-up, relative to the diabetic subjects with a large PEI (i.e., Group A) after adjustment for entry age, waist circumference, BMI, systolic and diastolic blood pressure, total cholesterol, triglyceride, pulse pressure, high-density lipoprotein cholesterol, low-density lipoprotein cholesterol, glycosylated hemoglobin, and fasting blood sugar. In addition, the Cox model revealed a graded association, with the diabetic subjects with a moderate PEI (i.e., Group B) having almost equal risks for developing DPN on follow-up relative to the diabetic subjects with a large PEI (i.e., Group A).

Table 4. Progression to DPN within six years of follow-up and relative risks as a function of three categories of PEI.

Categories of PEI Values	Subjects at Risk (n)	Events of DPN (n)	Relative Risk (95% CI)
Group A	22	6	1.00 (reference)
Group B	42	10	0.95 (0.63–2.07)
Group C	21	11	2.90 (1.58–6.87)
Total	85	27	—

Group A, diabetic subjects with a large PEI (the upper 25%); Group B, diabetic subjects with a moderate PEI (the middle 50%); Group C, diabetic subjects with a small PEI (the lower 25%). Relative risk estimated from a Cox proportional hazards survival model with adjustment for entry age, body mass index, resting systolic and diastolic blood pressure, total cholesterol, triglyceride, pulse pressure, high-density lipoprotein cholesterol, low-density lipoprotein cholesterol, and glycosylated hemoglobin. DPN, diabetic peripheral neuropathy; CI, confidence interval. The Cox proportional hazard survival model in SPSS was adopted. Events of DPN, number of future developing peripheral neuropathy in type 2 diabetic subjects within six years.

3.4. Cox Regression Analysis

The regression analysis using the Cox proportional hazards regression analysis of risk factors for incidence of DPN is shown in Table 5. The PEI was also significantly associated with the risk of developing DPN when it was treated as a continuous variable. The relative risk of incidence of DPN within six years of follow-up in diabetic patients for the PEI was 4.77 ($p = 0.015$), whereas the relative risks of incidence of DPN for the value of fasting plasma glucose and glycosylated hemoglobin were 1.01 ($p = 0.033$) and 0.73 ($p = 0.041$), respectively. The term of interaction between HbA1c and FPG was not significant ($p = 0.205$).

Compared with fasting plasma glucose and glycosylated hemoglobin, smaller PEI values can provide valid information that may help identify type 2 diabetic patients at a greater relative risk of future DPN from baseline measurement (i.e., PEI provided) to the end of the follow-up period (i.e., within six years after the PEI was provided).

Table 5. A proportional hazards analysis of risk factors for incidence of DPN within six years of follow-up in diabetic patients.

Risk Factors	Relative Risk	95% CI	*p* Values
HbA1c, %	0.73	0.52–1.05	0.041
FPG, mg/dL	1.01	1.00–1.02	0.033
HbA1c× FPG	1.00	1.00–1.01	0.205
PEI	4.77	1.87–6.31	0.015

HbA1c, glycosylated hemoglobin; FPG, fasting plasma glucose; PEI, percussion entropy index; DPN, diabetic peripheral neuropathy; CI, confidence interval. The variables of the models are entry age, body mass index, resting systolic and diastolic blood pressure, total cholesterol, triglyceride, pulse pressure, high-density lipoprotein cholesterol, low-density lipoprotein cholesterol, and glycosylated hemoglobin from baseline to the end of follow-up. Cox proportional hazards regression analysis in SPSS was adopted. A *p*-value < 0.05 was noted as statistically significant.

4. Discussion

Type 2 diabetes and its related complications are associated with the long-term damage and failure of various organ systems [31]. The overall impact of bad glucose control on vascular complications and major clinical outcomes in type 2 diabetes is still an open problem. While good glucose control has an undoubted benefit in the microvascular system of diabetic patients [6–8,32], good blood glucose control also improves microvascular disease and should be implemented early and maintained for the optimum length of time. A previous review study [31] highlighted the need for implementing programs for early detection, screening, and awareness to mitigate the burden of managing the complications. Therefore, diagnosis classifies a patient as having or not having a particular disease. In fact, diagnosis was recognized as the primary guideline for treatment and prognosis (i.e., what is going to happen in the future), and is still considered the key component of clinical practice [33]. However, it is not an easy job to control the appropriate glucose for diabetic patients; this is the reason DPN is still one of the most common chronic complications of diabetes. Recently, a study [23] that determined the exact PPI intervals was reported to provide a prognosis of peripheral neuropathy from diabetic patients. However, real-time computation was not able to obtain immediate index information of the test subjects. Furthermore, it did not provide valid information regarding the degree of risk of developing future DPN that may encourage type 2 diabetic patients to follow a good lifestyle.

This study addressed results from the indices LHR, MEI_{SS}, MEI_{LS}, PWV_{mean}, and PEI, which were first computed for diabetic subjects with peripheral neuropathy six years after baseline measurement (i.e., Group 3) for comparison with diabetic patients without peripheral neuropathy (i.e., Group 2). Although the values of the vascular stiffness indices, including MEI_{LS} and PWV_{mean}, were significantly different in Group 2 compared with the Group 1 subjects ($p < 0.017$), there were no notable differences between Groups 2 and 3 ($p > 0.017$). On the other hand, the PEI (i.e., BRS assessment index) showed highly significant differences among the three groups ($p < 0.017$) (Table 2). These results are consistent with the major outcomes in the previous study [23]. Significantly smaller values for the PEI were noted for Group 3 compared to the other two groups (e.g., Group 1 vs. Group 2 vs. Group 3: 0.73 ± 0.01 vs. 0.63 ± 0.01 vs. 0.59 ± 0.02), which is consistent with the same findings, where diabetic neuropathy was found to be a more significant crucial factor of spontaneous BRS assessment than carotid elasticity in type 2 diabetics in [34].

Therefore, most diabetic patients with a smaller PEI value were only concerned about the relative risks of the future development of DPN, and were not focused on achieving a smaller PEI value, because DPN, which has a lifetime prevalence of approximately 50%, is the most common diabetic complication. DPN is also the leading cause of disability due to diabetic foot ulceration and amputation, gait disturbance, and fall-related injury [3,35]. Neuropathy not only causes problems such as a decreased quality of life, poor sleep, and depression in diabetic patients, but the quality of life is also greatly affected [36–38]. Although PEI has recently been introduced to assess the complexity of BRS [21–23], the significance of smaller PEI values concerning the identification of subjects with type 2 diabetes who are more prone to develop diabetic neuropathy is unknown. That is, it would be difficult to predict how many and who will develop DPN in advance. Thus, a model of clinical practice focused on DPN prognosis and predicting the likelihood of future outcomes associated with PEI may be more useful for diabetic patients [33].

According to the results in Table 2, the present study adopted the PEI as the first measurement of all of the recruited diabetic patients to create the basis of quartiles in the diabetic population's distribution of the PEI: the upper 25% (i.e., $n = 22$, Group A, 6 DPN included), the middle 50% (i.e., $n = 42$, Group B, 10 DPN included), and the lower 25% (i.e., $n = 21$, Group C, 11 DPN included). The diabetic patients in Group C showed remarkably higher HbA1c levels than those in the diabetic patients in Group B ($p < 0.017$) (Table 3). On the other hand, no significant differences were noted in the demographic and hemodynamic parameters, as well as the fasting blood glucose and serum lipid profile between any two groups (i.e., Group A vs. Group B and Group B vs. Group C) (Table 3). In the diabetic patients with smaller PEI values in Group 3, almost 50% had developed DPN within six years (Table 4). These results are consistent with statements in the study [31]. According to goodness-of-fit test result in the study, the null hypothesis (i.e., no association between smaller PEI values and developing DPN within six years after the PEI was provided) was rejected, with the chi-square value being larger than the level of significance. An association exists between diabetic patients with smaller PEI values and the development of DPN within six years after baseline measurement.

A total of 27 type 2 diabetic patients developed DPN among 85 study patients (31.8%) in the six years after baseline examinations. This finding is consistent with similar results reported in [37,38]. The progression to DPN of patients in the three categories within six years and the corresponding relative risks for the incidence of DPN assessed by the Cox proportional hazard survival model are shown in Table 4. The Cox model revealed a graded association, with the diabetic subjects with a small PEI (i.e., Group C) at 2.90-times greater risk of developing DPN on follow-up relative to the diabetic subjects with a large PEI (i.e., Group A) after adjustment for entry age, waist circumference, BMI, systolic blood pressure and diastolic blood pressure, total cholesterol, triglyceride, pulse pressure, high-density lipoprotein cholesterol, low-density lipoprotein cholesterol, glycosylated hemoglobin, and fasting plasma glucose. Type 2 diabetic patients with smaller PEI values may have a larger relative risk of developing DPN within six years. In addition, the relative risk of incidence of DPN within six years of follow-up in diabetic patients for the PEI was 4.77 ($p = 0.015$) in Table 5. This study was designed to use synchronized ECG and PPG signals (i.e., PEI) in predicting the development of peripheral neuropathy from type 2 diabetes. The PEI has recently been introduced not only to assess the complexity of BRS but also to show the significance of smaller PEI values concerning the identification of subjects with type 2 diabetes who are more prone to DPN.

The current study has its limitations. Firstly, the number of subjects recruited was relatively small. In addition, it may be difference between DPN attack confirmation time and checkout time in our study for non-fixed follow-up to each diabetic patient. Therefore, Kaplan–Meier survival analysis was not adopted in this study. Nevertheless, highly significant associations between PEIs and relative risks of developing DPN were still significant. Secondly, this study focused on the Cox proportional hazard survival model for diabetic patients, and the optimal BRS delay between the amplitude and RRI series would be set at one to five heartbeat cycles for all test subjects with the same setting. Thirdly, the impact of periodontal therapy on diabetes control was not investigated because of the limited number of diabetic patients. Subsequently, the period of baseline measurement was more than two years, because of the limited number of subjects in each group. Finally, as an observational study, the values of PEI and the proposed parameters could be used to identify the risk factors for a prediction task by using simple machine learning algorithms (such as SVM, LDA, or even deep learning) in the future.

5. Conclusions

This study represents the first attempt to investigate the clinical prognostic feasibility of applying the PEI, and demonstrates enhanced sensitivity in differentiating between diabetic subjects without DPN and diabetic subjects with DPN within six years after baseline measurement, compared to single-scale indices (i.e., LHR and PWV_{mean}) and multiple temporal-scale indices (i.e., MEI_{SS} and MEI_{LS}). Our findings suggest that diabetic patients with smaller PEI values are more prone to developing DPN, which is of potential importance for application in the area of the point-of-care diagnostic devices.

Author Contributions: Conceptualization, H.-T.W.; Data curation, N.T., W.-R.H., S.-Y.W., and J.-J.C.; Formal analysis, N.T. and H.-T.W.; Funding acquisition, H.-C.W.; Investigation, N.T., M.-X.X., X.-J.T., and H.-T.W.; Methodology, N.T., J.-J.C., and H.-T.W.; Project administration, H.-C.W. and H.-T.W.; Resources, H.-C.W.; Software, N.T., W.-R.H., S.-Y.W., and M.-X.X.; and Supervision, H.-C.W., X.-J.T., and H.-T.W. All authors have read and agreed to the published version of the manuscript.

Funding: This research was funded by North Minzu University Scientific Research Projects (Major projects No. 2019KJ37), National Natural Science Foundation of China (No. 61861001), Ningxia Municipal Health Commission Project (No. 2018NW007), and "Tian Cheng Hui Zhi" Innovation & Education Fund of Chinese Ministry of Education (No. 2018A01016).

Acknowledgments: The authors are grateful for the support of Texas Instruments, Taiwan, in sponsoring the MSP tools and assisting in developing novel signal-processing techniques as a contribution to preventive medicine in this study. The authors would like to thank the Guest editor and Reviews for their insightful recommendations, which have contributed greatly to the improvement of this work.

Conflicts of Interest: The authors declare no conflict of interest.

Abbreviations

BMI	Body Mass Index
BRS	Baroreflex Sensitivity
CI	Confidence Interval
DBP	Diastolic Blood Pressure
DM	Diabetes Mellitus
DPN	Diabetic Peripheral Neuropathy
DVP	Digital Volume Pulse
ECG	Electrocardiography
FPG	Fasting Plasma Glucose
HbA1c	Glycosylated hemoglobin
HDL	High-Density Lipoprotein cholesterol
HFP	High Frequency Power
HRV	Heart Rate Variability
LDL	Low-Density Lipoprotein cholesterol
LFP	Low-Frequency Power
LHR	Low-/High-frequency power Ratio (LFP/HFP, LHR) using the RRI dataset
MEI	Multiscale Entropy Index using the RRI dataset only
PEI	Percussion Entropy Index using synchronized {RRI} and {Amp} signals
PN	Peripheral Neuropathy
PP	Pulse Pressure

PPG	Photoplethysmography
PPI	Peak-to-Peak Interval
PWV	Pulse Wave Velocity
RR	Relative Risks
RRI	R-R Interval of ECG
SBP	Systolic Blood Pressure
SD	Standard Deviation
SPSS	Statistical Package for the Social Sciences
TG	Triglyceride
WC	Waist Circumference

References

1. Marshall, S.M.; Flyvbjerg, A. Prevention and early detection of vascular complications of diabetes. *BMJ* **2006**, *333*, 475–480. [CrossRef] [PubMed]
2. Beckman, J.A.; Creager, M.A. Vascular complications of diabetes. *Circ. Res.* **2016**, *118*, 1771–1785. [CrossRef] [PubMed]
3. Juster-Switlyk, K.; Smith, A.G. Updates in diabetic peripheral neuropathy. *F1000Research* **2016**, *5*. [CrossRef] [PubMed]
4. Iqbal, Z.; Azmi, S.; Yadav, R.; Ferdousi, M.; Kumar, M.; Cuthbertson, D.J.; Lim, J.; Malik, R.A.; Alam, U. Diabetic peripheral neuropathy: Epidemiology, diagnosis, and pharmacotherapy. *Clin. Ther.* **2018**, *40*, 828–849. [CrossRef] [PubMed]
5. Vaidya, V.; Gangan, N.; Sheehan, J. Impact of cardiovascular complications among patients with Type 2 diabetes mellitus: A systematic review. *Expert Rev. Pharmacoecon. Outcomes Res.* **2015**, *15*, 487–497. [CrossRef]
6. Lin, I.W.; Chang, H.H.; Lee, Y.H.; Wu, Y.C.; Lu, C.W.; Huang, K.C. Blood sugar control among type 2 diabetic patients who travel abroad: A cross sectional study. *Medicine* **2019**, *98*, e14946. [CrossRef]
7. Home, P. Safety of very tight blood glucose control in type 2 diabetes. *BMJ* **2008**, *336*, 458–459. [CrossRef]
8. Balkau, B.; Calvi-Gries, F.; Freemantle, N.; Vincent, M.; Pilorget, V.; Home, P.D. Predictors of HbA1c over 4 years in people with type 2 diabetes starting insulin therapies: The CREDIT study. *Diabetes Res. Clin. Pract.* **2015**, *108*, 432–440. [CrossRef]
9. Nayak, S.; Blumenfeld, N.R.; Laksanasopin, T.; Sia, S.K. Point-of-care diagnostics: Recent developments in a connected age. *Anal. Chem.* **2017**, *89*, 102–123. [CrossRef]
10. Bonetti, P.O.; Lerman, L.O.; Lerman, A. Endothelial dysfunction: A marker of atherosclerotic risk. *Arterioscler. Thromb. Vasc. Biol.* **2003**, *23*, 168–175. [CrossRef]
11. Bonetti, P.O.; Pumper, G.M.; Higano, S.T.; Holmes, D.R., Jr.; Kuvin, J.T.; Lerman, A. Noninvasive identification of patients with early coronary atherosclerosis by assessment of digital reactive hyperemia. *J. Am. Coll. Cardiol.* **2004**, *44*, 2137–2141. [CrossRef] [PubMed]
12. Grover-Paez, F.; Zavalza-Gomez, A.B. Endothelial dysfunction and cardiovascular risk factors. *Diabetes Res. Clin. Pract.* **2009**, *84*, 1–10. [CrossRef] [PubMed]
13. Quattrini, C.; Jeziorska, M.; Boulton, A.J.; Malik, R.A. Reduced vascular endothelial growth factor expression and intra-epidermal nerve fiber loss in human diabetic neuropathy. *Diabetes Care* **2008**, *31*, 140–145. [CrossRef] [PubMed]
14. Malik, M.; Bigger, J.T.; Camm, A.J.; Kleiger, R.E. Heart rate variability. Standards of measurement, physiological interpretation, and clinical use. Task Force of the European Society of Cardiology and the North American Society of Pacing and Electrophysiology. *Eur. Heart J.* **1996**, *17*, 354–381. [CrossRef]
15. Pozza, R.D.; Bechtold, S.; Bonfig, W.; Putzker, S.; Kozlik-Feldmann, R.; Schwarz, H.-P.; Netz, H. Impaired short-term blood pressure regulation and autonomic dysbalance in children with type 1 diabetes mellitus. *Diabetologia* **2007**, *50*, 2417–2423. [CrossRef]
16. Rosengard-Barlund, M.; Bernardi, L.; Fagerudd, J.; Mantysaari, M.; Af Bjorkesten, C.G.; Lindholm, H.; Forsblom, C.; Waden, J.; Groop, P.H. Early autonomic dysfunction in type 1 diabetes: A reversible disorder? *Diabetologia* **2009**, *52*, 1164–1172. [CrossRef]

17. Yuan, H.K.; Lin, C.; Tsai, P.H.; Chang, F.C.; Lin, K.P.; Hu, H.H.; Su, M.C.; Lo, M.T. Acute increase of complexity in the neuro-cardiovascular dynamics following carotid stenting. *Acta Neurol. Scand.* **2011**, *123*, 187–192. [CrossRef]
18. Lerma, C.; Infante, O.; Perez-Grovas, H.; Jose, M.V. Poincare plot indexes of heart rate variability capture dynamic adaptations after haemodialysis in chronic renal failure patients. *Clin. Physiol. Funct. Imaging* **2003**, *23*, 72–80.
19. Merati, G.; Di Rienzo, M.; Parati, G.; Veicsteinas, A.; Castiglioni, P. Assessment of the autonomic control of heart rate variability in healthy and spinal-cord injured subjects: Contribution of different complexity-based estimators. *IEEE Trans. Biomed. Eng.* **2006**, *53*, 43–52. [CrossRef]
20. Pan, W.Y.; Su, M.C.; Wu, H.T.; Su, T.J.; Lin, M.C.; Sun, C.K. Multiscale entropic assessment of autonomic dysfunction in patients with obstructive sleep apnea and therapeutic impact of continuous positive airway pressure treatment. *Sleep Med.* **2016**, *20*, 12–17. [CrossRef]
21. Wei, H.C.; Xiao, M.X.; Ta, N.; Wu, H.T.; Sun, C.K. Assessment of diabetic autonomic nervous dysfunction with a novel percussion entropy approach. *Complexity* **2019**, *2019*, 6469853. [CrossRef]
22. Xiao, M.X.; Lu, C.H.; Ta, N.; Jiang, W.W.; Tang, X.J.; Wu, H.T. Application of a Speedy Modified Entropy Method in Assessing the Complexity of Baroreflex Sensitivity for Age-Controlled Healthy and Diabetic Subjects. *Entropy* **2019**, *21*, 894. [CrossRef]
23. Wei, H.-C.; Ta, N.; Hu, W.-R.; Xiao, M.-X.; Tang, X.-J.; Haryadi, B.; Liou, J.J.; Wu, H.-T. Digital Volume Pulse Measured at the Fingertip as an Indicator of Diabetic Peripheral Neuropathy in the Aged and Diabetic. *Entropy* **2019**, *21*, 1229. [CrossRef]
24. Zhang, M.; Bai, Y.; Ye, P.; Luo, L.; Xiao, W.; Wu, H.; Liu, D. Type 2 diabetes is associated with increased pulse wave velocity measured at different sites of the arterial system but not augmentation index in a Chinese population. *Clin. Cardiol.* **2011**, *34*, 622–627. [CrossRef] [PubMed]
25. Tsai, W.C.; Chen, J.Y.; Wang, M.C.; Wu, H.T.; Chi, C.K.; Chen, Y.K.; Chen, J.H.; Lin, L.J. Association of Risk Factors With Increased Pulse Wave Velocity Detected by a Novel Method Using Dual-Channel Photoplethysmography. *Am. J. Hypertens.* **2005**, *18*, 1118–1122. [CrossRef]
26. Wu, H.T.; Hsu, P.C.; Liu, A.B.; Chen, Z.L.; Huang, R.M.; Chen, C.P.; Tang, C.J.; Sun, C.K. Six-channel ECG-based pulse wave velocity for assessing whole-body arterial stiffness. *Blood Press.* **2012**, *21*, 167–176. [CrossRef]
27. Wu, H.T.; Hsu, P.C.; Lin, C.F.; Wang, H.J.; Sun, C.K.; Liu, A.B.; Lo, M.T.; Tang, C.J. Multiscale entropy analysis of pulse wave velocity for assessing atherosclerosis in the aged and diabetic. *IEEE Trans. Biomed. Eng.* **2011**, *58*, 2978–2981.
28. American Diabetes Association. Diagnosis and classification of diabetes mellitus. *Diabetes Care* **2014**, *37* (Suppl. 1), S81–S90. [CrossRef]
29. Jin, H.Y.; Lee, K.A.; Park, T.S. The impact of glycemic variability on diabetic peripheral neuropathy. *Endocrine* **2016**, *53*, 643–648. [CrossRef]
30. Harrell, F.E., Jr. (Ed.) "Introduction to survival analysis," and "Parametric survival models". In *Regression Modeling Strategies with Applications to Linear Models, Logistic and Ordinal Regression, and Survival Analysis*, 2nd ed.; Springer Series in Statistics: New York, NY, USA, 2015; pp. 399–451.
31. Chawla, A.; Chawla, R.; Jaggi, S. Microvasular and macrovascular complications in diabetes mellitus: Distinct or continuum? *Indian J. Endocrinol. Metab.* **2016**, *20*, 546–551. [CrossRef]
32. Giorgino, F.; Home, P.D.; Tuomilehto, J. Glucose control and vascular outcomes in Type 2 diabetes: Is the picture clear? *Diabetes Care* **2016**, *39* (Suppl. 2), S187–S195. [CrossRef]
33. Croft, P.; Altman, D.G.; Deeks, J.J.; Dunn, K.M.; Hay, A.D.; Hemingway, H.; Timmis, A. The science of clinical practice: Disease diagnosis or patient prognosis? Evidence about "what is likely to happen" should shape clinical practice. *BMC Med.* **2015**, *13*, 20. [CrossRef]
34. Ruiz, J.; Monbaron, D.; Parati, G.; Perret, S.; Haesler, E.; Danzeisen, C.; Hayoz, D. Diabetic neuropathy is a more important determinant of baroreflex sensitivity than carotid elasticity in type 2 diabetes. *Hypertension* **2005**, *46*, 162–167. [CrossRef] [PubMed]
35. Colberg, S.R.; Sigal, R.J.; Yardley, J.E.; Riddell, M.C.; Dunstan, D.W.; Dempsey, P.C.; Horton, E.S.; Castorino, K.; Tate, D.F. Physical activity/exercise and diabetes: A position statement of the American Diabetes Association. *Diabetes Care* **2016**, *39*, 2065–2079. [CrossRef] [PubMed]

36. Gonzalez-Martin, C.; Pertega-Diaz, S.; Seoane-Pillado, T.; Balboa-Barreiro, V.; Soto-Gonzalez, A.; Veiga-Seijo, R. Structural, Dermal and Ungual Characteristics of the Foot in Patients with Type II Diabetes. *Medicina* **2019**, *55*, 639. [CrossRef]
37. Vinik, A.; Casellini, C.; Nevoret, M.L. Diabetic Neuropathies. In *Endotext [Internet]*; Feingold, K.R., Anawalt, B., Boyce, A., Eds.; South Dartmouth (MA): MDText.com, Inc.: Boston, MA, USA, 2000. Available online: https://www.ncbi.nlm.nih.gov/books/NBK279175/ (accessed on 9 November 2019).
38. Schreiber, A.K.; Nones, C.F.; Reis, R.C.; Chichorro, J.G.; Cunha, J.M. Diabetic neuropathic pain: Physiopathology and treatment. *World J. Diabetes* **2015**, *6*, 432–444. [CrossRef] [PubMed]

Article

Blood Ketone Bodies and Breath Acetone Analysis and Their Correlations in Type 2 Diabetes Mellitus

Valentine Saasa [1,2,*], Mervyn Beukes [3], Yolandy Lemmer [4] and Bonex Mwakikunga [1,5,*]

1 DSI/CSIR Centre for Nanostructures and Advanced Materials, P.O. Box 3951, Pretoria 0001, South Africa
2 Department of Biochemistry, Genetics and Microbiology, University of Pretoria, Pretoria 0001, South Africa
3 Department of Biochemistry, Stellenbosch University, Cape Town 7600, South Africa; mervynbeukes@sun.ac.za
4 Next Generation Health, Division 1, CSIR, P.O. Box 3951 Pretoria, South Africa; ylemmer@csir.co.za
5 Department of Physics, Tshwane University of Technology, P.O. Box x680, Pretoria 0001, South Africa
* Correspondence: vsaasa@csir.co.za (V.S.); bmwakikunga@csir.co.za (B.M.); Tel.: +27-12-841-3601 (V.S.); +27-12-841-4771 (B.M.)

Received: 4 November 2019; Accepted: 7 December 2019; Published: 17 December 2019

Abstract: Analysis of volatile organic compounds in the breath for disease detection and monitoring has gained momentum and clinical significance due to its rapid test results and non-invasiveness, especially for diabetes mellitus (DM). Studies have suggested that breath gases, including acetone, may be related to simultaneous blood glucose (BG) and blood ketone levels in adults with types 2 and 1 diabetes. Detecting altered concentrations of ketones in the breath, blood and urine may be crucial for the diagnosis and monitoring of diabetes mellitus. This study assesses the efficacy of a simple breath test as a non-invasive means of diabetes monitoring in adults with type 2 diabetes mellitus. Human breath samples were collected in Tedlar™ bags and analyzed by headspace solid-phase microextraction and gas chromatography-mass spectrometry (HS-SPME/GC-MS). The measurements were compared with capillary BG and blood ketone levels (β-hydroxybutyrate and acetoacetate) taken at the same time on a single visit to a routine hospital clinic in 30 subjects with type 2 diabetes and 28 control volunteers. Ketone bodies of diabetic subjects showed a significant increase when compared to the control subjects; however, the ketone levels were was controlled in both diabetic and non-diabetic volunteers. Worthy of note, a statistically significant relationship was found between breath acetone and blood acetoacetate ($R = 0.89$) and between breath acetone and β-hydroxybutyrate ($R = 0.82$).

Keywords: diabetes mellitus; ketone bodies; human breath; acetone; beta-hydroxybutyrate; acetoacetate; gas chromatography-mass spectrometry (GC-MS)

1. Introduction

Human biological samples such as breath, blood and urine contain several volatile organic compounds (VOCs). These VOCs are associated with specific metabolic pathways, and are useful as biomarkers reflecting the disease and physiological state of a human that cause changes in their metabolism [1]. Particularly, analysis of breath has been receiving more attention because of its potential as a non-invasive method for disease diagnosis and metabolic status monitoring [2]. Among thousands of VOCs, acetone is the second to highest in abundance in normal human breath gases, which has been extensively studied as a breath biomarker of diabetes or as a high abundant breath VOC in various physiological cases since the 1950s [1,3]. The studies which showed a strong link between breath acetone and plasma glucose are mostly for type 1 diabetes, but no such observation has been obtained so far from adequately controlled type 2 diabetes mellitus patients [4–7].

In addition, breath acetone concentration (BrAce) is also well understood to be a non-invasive measure of ketosis. Ketosis is a metabolic state characterized by the elevation of ketone bodies in the blood. Healthy individuals on standard mixed diets (i.e., moderate to high carbohydrate content) have basal ketosis, while individuals with uncontrolled diabetes have extremely elevated ketosis, which could lead to ketoacidosis. In all cases, ketosis describes the quantity of circulating ketone bodies [8].

Ketone bodies are produced as a by-product of the fat metabolism process [9]. When the liver metabolizes circulating free fatty acids, these acids are transformed into acetyl-coenzyme A (acetyl-CoA), a molecule used in the production of energy. The acetyl-CoA linked to the tricarboxylic acid cycle (TCA cycle) is produced from glucose and fatty acids (Figure 1). Before acetyl CoA enters the TCA cycle, it first condenses with oxaloacetate, and when the glucose availability is reduced in the liver due to: fasting, a low-carbohydrate diet, insulin deficiency or insulin resistance caused by diabetes, the production of acetyl CoA from fatty acids increases. The resultant excess acetyl CoA is diverted to the production of ketone bodies. As a result, blood and breath acetone concentration increase [10]. Indeed, elevated acetone concentrations within exhaled breath have been observed in diabetes mellitus patients [6,11–14].

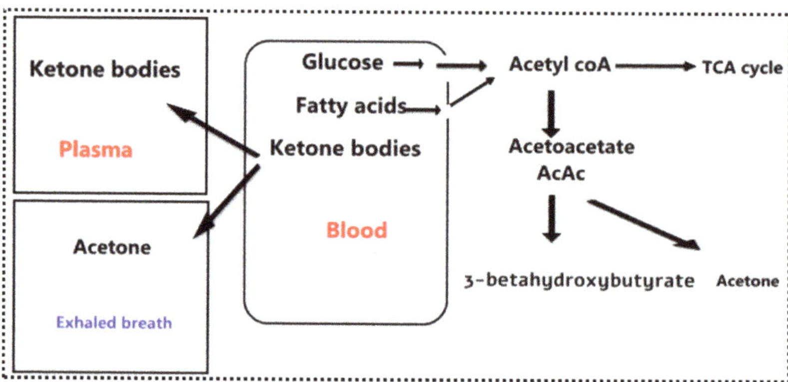

Figure 1. Schematic diagram of the formation of ketone bodies (acetoacetate, beta-hydroxybutyrate, and acetone) which takes place in the mitochondrial matrix of the liver cell. The acetyl-coenzyme A (acetyl-CoA) can be metabolized through the tricarboxylic acid (TCA) cycle, or can undergo ketogenesis. The three ketone bodies travel through the blood and acetone is also expelled the breath. (Red is for ketones in the blood and blue is for ketones in the human breath).

The world prevalence of diabetes mellitus among adults (aged 20–79 years) in 2010 was 6.4%, affecting 285 million adults, and is estimated to increase to 7.7%, affecting 439 million adults by 2030. Between 2010 and 2030, there will be a 69% increase in the numbers of adults with diabetes in developing countries, and a 20% increase in developed countries [15]. South Africa, as one of the developing countries, has half a million people (about 6% of the population) suffering from diabetes mellitus, the majority having type 2 diabetes mellitus [16,17]. The major abnormalities in type 2 diabetes mellitus include dyslipidemia, insulin resistance, hyperglycemia and ketoacidosis, which usually are not taken into consideration in type 2 diabetes mellitus [18]. Additionally, the diagnosis and monitoring of blood glucose and ketone bodies for diabetic patients involves the use of blood tests. Usually, this is done by drawing blood from a patient for analysis and using a glucose measuring device such as a glucometer. These methods are painful, invasive and time-consuming [19]. A great opportunity lies in the use of breath acetone for diagnosis and monitoring of the disease. This might suggest the potential to develop breath gas analysis to provide an alternative to blood testing for glucose and ketone measurement. This could offer patients a non-invasive, pain free and point of care solution to their daily lives.

Studies have suggested that breath gases, including acetone, may be related to simultaneous blood glucose (BG) and blood ketone levels in type 1 diabetes [20–22]. However, from our knowledge based on literature, there has not been a study on the analysis or correlation of ketone bodies and breath acetone on type 2 diabetes mellitus. In this study, we assessed the relationship between blood ketone bodies and breath acetone, along with the blood glucose and breath acetone of type 2 diabetes mellitus, and determined whether breath acetone could be used as a biomarker for diabetes mellitus. The aim of the present study is to establish the relationship between breath acetone and plasma ketones of non-diabetes mellitus volunteers and type 2 diabetes mellitus patients.

While breath acetone and blood ketone bodies have been measured in relatively large cohorts of diabetes mellitus patients, most breath and blood ketone measurements have been carried out on only type 1 diabetes mellitus. For example, Blaikie et al. [20] measured breath acetone and blood ketone bodies in children with type 1 diabetes mellitus, Minh et al. [23] showed that concentrations of two separate groups of acetone, methyl nitrate, ethanol and methylbenzene and acetone, methanol, propane and 2-pentyl nitrate were able to demonstrate that blood glucose can be measured simultaneously with breath acetone in both healthy adults with type 1 diabetes mellitus. A previous study from the same group demonstrated a correlation between exhaled methyl nitrate and blood glucose in a cohort of young people with type 1 diabetes mellitus [24]. Many more studies on breath acetone, blood ketone and blood glucose have been reported [7,25,26]. Nevertheless, the urge to look for the non-invasive monitoring of both blood glucose and ketone bodies should not be limited to only type 1 diabetes mellitus, as type 2 diabetic patients can also suffer from ketoacidosis, and this therefore also requires daily monitoring of their blood glucose non-invasively.

The aim of this study is to assess the efficacy of a simple breath test as a non-invasive means of diabetes monitoring in type 2 diabetes mellitus patients. Human breath samples were collected in Tedlar™ bags and analyzed by HS-SPME/GC-MS. The measurements were compared with capillary fasting blood glucose (BG) and ketone levels taken at the same time on a single visit to a routine hospital clinic in 30 subjects with type 2 diabetes mellitus and 28 control volunteers. A statistically significant relationship was found between breath acetone and blood acetoacetate (R = 0.897), and between breath acetone and β-hydroxybutyrate (R = 0.821).

2. Materials and Methods

2.1. Study Population

A total of thirty (30) diabetes mellitus and twenty-eight (28) controls participants aged between 18 to 60 years were recruited for this study based on the prevalence of diabetes in this locality. The patients were all non-smokers. Inclusion criteria include being diabetic, while an exclusion criterion involved being non-diabetic and having any other chronic illness. The fasting blood was used. A routine glucose test was also performed for participants to confirm the diabetic state. Informed consent was obtained from participants as well as ethical clearance (Ref: 118/2015, 21 November 2016) from the ethics committee of the Council for Scientific and Industrial Research (CSIR).

2.2. Collection of Samples

Blood samples were collected by standard vein puncture into the plain tube in the morning before a meal (after 8 h since the last meal). The blood was centrifuged at 3000 rpm for 10 min. The serum was separated into a separate tube and kept in −80 °C freezer until analysis. The breath samples were also collected simultaneously with blood samples using the Tedlar™ bags via a two-way non-rebreathing valve and analyzed immediately using HS-SPME/GC-MS. The participants were asked to inhale moderately and then to exhale as much as possible into a 3-lTedlar™ bag. Tedlar™ bags were first flushed with with pure nitrogen gas prior to the collection of breath samples.

2.3. Biochemical Analysis in the Blood

Ketone bodies (acetoacetate and beta-hydroxybutyrate), were analyzed using Abcam's Acetoacetate (ab180875) and Beta-hydroxybutyrate (ab83390), respectively. The Acetoacetate kits provide a sensitive method to quantitate endogenous levels of AcAc in human blood. In this non-enzymatic assay, AcAc reacts with a substrate to generate a colored product that can be measured at 550 nm [27]. The reaction is specific for AcAc, and does not detect beta-hydroxybutyrate. The beta-hydroxybutyrate kits utilize beta HB dehydrogenase to generate a product that reacts with the colorimetric probe with an absorbance peak of 450 nm [28].

Serum from 30 diabetic and 28 non-diabetic mellitus patients were analyzed using Abcam's glucose assay kit to quantify the amount of glucose in the blood. In this assay, glucose oxidase specifically oxidizes glucose to generate a product which reacts with a dye to generate color (570 nm). The generated color is proportional to the glucose amount.

2.4. Breath Acetone Analysis Using HS-SPME/GC-MS

In this study, we used the HS-SPME/GC-MS to quantify the breath acetone in diabetic and non-diabetic mellitus patients. The method is very simple, fast and sensitive for the optimization and accurate quantification of acetone in human breath. Acetone gas standards in the concentration range of 0.073, 0.59, 1.66, 3.32 and 6.64 ppmv were prepared. A solid-phase microextraction (SPME) fiber pre-coated with 20 mg/mL of *O*-2,3,4,5,6-(pentafluorobenzyl) hydroxylamine hydrochloride (PFBHA) was exposed inside the 3 L Tedlar™ bag at 40 °C containing the acetone standards and human breath for 10 min. Acetone present in the breath samples underwent on-fiber SPME derivatization to form the stable oxime (Figure 2).

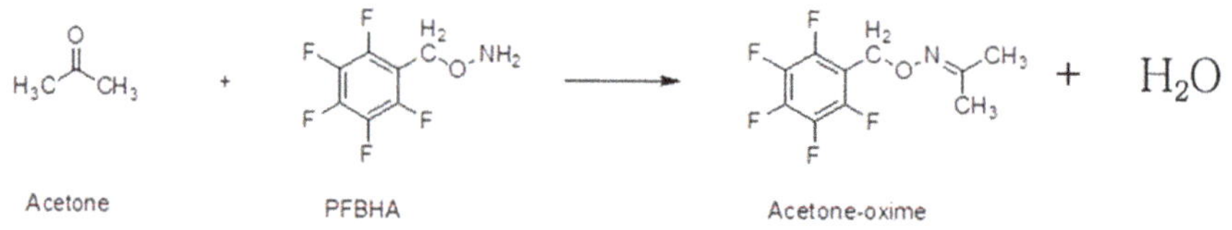

$$\text{Acetone} + \text{PFBHA} \longrightarrow \text{Acetone-oxime} + H_2O$$

Figure 2. Schematic representation of the reaction between breath acetone and the derivatizing agent (PFBHA) reacting on the solid-phase microextraction (SPME) fiber.

An Agilent Technologies model 6890N Gas Chromatography interfaced with Agilent Technologies model 6890N Mass Selective Detector was used for analysis. A 30 m × 0.25 mm with 0.25 µm RXi®-5 SilMS (Restek, Bellefonte, Pennsylvania, USA) was used as the analytical column. The injection port temperature which was also the temperature for thermal desorption of VOC was 250 °C, and the desorption time was 2 min. The GC split valve was set to open after the 2 min desorption Btime. The GC injector liner was a quartz SPME liner with an internal diameter of 0.75 mm (Supelco Inc., Bellefonte, PA, USA). The column temperature was regulated through the program, an initial temperature of 60 °C, was increased to 150 °C at 10 °C /min, and then increased to 300 °C (and held for 1 min). Total ion current was monitored using the electron-impact ionization mode (70 EV). Quantification was performed using characteristic mass. The peak at *m*/*z* 181 was used for the quantification of the acetone-PFBHA derivative.

2.5. Statistical Analysis

Statistical Software Package for Social Sciences (SPSS) 26 was used for data processing. Results are expressed in mean ± standard deviation (SD) or median (range). The correlation studies were investigated using Pearson and Spearman's rank correlations. Linear models were then fitted with the breath acetone concentration as response variables. The data for qualitative comparison was analyzed by using Levene's *t*-test. A p-value < 0.05 was considered statistically significant.

3. Results

3.1. Biochemical Analysis

Assessment of ketone bodies is an important practice more especially in clinical institutions to check and monitor for diabetic ketoacidosis (DKA) and to ascertain whether breath acetone can be used in monitoring and controlling the disease. The high amount of ketone bodies in the blood is usually an indicator that the catabolism of fatty acid is greater than the one for carbohydrates [29,30]. In this study, 30 diabetic and 28 non-diabetic patients' fasting serums were analyzed for ketone bodies (acetoacetate and beta-hydroxybutyrate) using the Abcam® acetoacetate and beta-hydroxybutyrate assay kits (Figure 3). Other clinical data, which include blood glucose, glycated hemoglobin, total cholesterol, triglycerides, high-density lipoprotein and low-density lipoprotein were also measured to confirm the state of diabetes mellitus (Table 1).

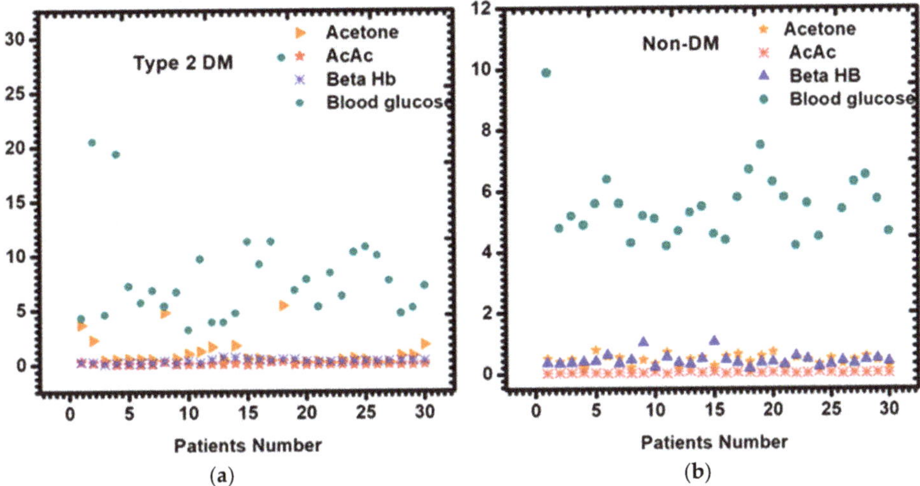

Figure 3. Scatter plot for plasma blood glucose, acetoacetate, beta-hydroxybutyrate and breath acetone in (**a**) type 2 diabetic and (**b**) non-diabetic mellitus patients.

Table 1. Clinical data of type 2 diabetes mellitus (DM) and non-diabetes mellitus.

Biochemical Parameters	Type 2 DM (n = 30)	Non-Diabetes (n = 30)	p-Value
Age	47 ± 10	41 ± 10	<0.001
Gender	13/17	11/19	0.10
BMI (kg·m^{-2})	28.4 ± 4.5	25.4 ± 4.0	0.47
Plasma glucose (mmol/L)	8.6 ± 2.43	5.7 ± 1.44	0.007
HB1Ac (%)	10.3 ± 2.57	-	-
Total cholesterol (mmol/L)	5.10 ± 1.40	4.5 ± 1.42	0.17
Triglycerides (mmol/L)	1.57 ± 1.3	1.04 ± 1	0.01
HDL cholesterol (mmol/L)	1.15 ± 0.27	1.33 ± 0.47	0.34
LDL cholesterol (mmol/L)	2.56 ± 1.32	2.43 ± 0.97	0.52
B-hydroxybutyrate (mmol/L)	0.46 ± 0.02	0.44 ± 0.41	0.55
Acetoacetate (mmol/L)	0.09 ± 0.02	0.05 ± 0.03	0.47

Data in mean ± standard deviation (SD), BMI (body mass index), HDL (high-density lipoprotein) and LDL (low-density lipoprotein).

Breath acetone from both diabetic and non-diabetic patients was analyzed with the HS-SPME/GC-MS. As expected, guided by literature, a high amount of breath acetone was observed in diabetic patients (more than 0.8 ppm), as opposed to their non-diabetic (less than 0.8 ppm) counterparts.

The mean values of acetoacetate in diabetic and non-diabetic patients were 0.09 mmol/L and 0.05 mmol/L, respectively. The mean β-hydroxybutyrate was also higher in diabetes patients, 0.46 mmol/L), as compared to the non-diabetes patients 0.25 mmol/L. As shown in Table 1, plasma glucose and total cholesterol were slightly higher in type 2 diabetes mellitus than non-diabetes mellitus. Triglycerides, HDL cholesterol, LDL cholesterol, β-hydroxybutyrate and acetoacetate levels were not significantly different between the two groups.

3.2. Breath Acetone Analysis Using HS-SPME/GC-MS

In this study, HS-SPME/GC-MS analysis was applied to determine the concentration of acetone in the breath of 30 diabetic and 28 non-diabetic mellitus patients. Given the small mass of acetone (58 amu) and its volatility, the acetone in breath was first derivatized with a derivatizing reagent, pentafluorobenzyl-hydroxylamine-hydrochloride (PFBHA) prior to the GC-MS analysis. The reaction of acetone in the breath with a derivatizing agent (PFBHA) forms a very stable acetone-oxime, and Figure 4 shows the mass spectrum of the acetone oxime with a base peak at *m/z* 181 which was extracted for quantitative purposes. The amount of oxime formed on the fiber is proportional to the acetone concentration in the breath. Acetone concentration higher than 1.8 ppm was observed in some of the diabetic breath (Figure 3a) and acetone concentration lower than 0.8 ppm was observed (Figure 3b). The GC spectrum of both diabetic and non-diabetic mellitus volunteers are available on the supplementarary results in Figure S1.

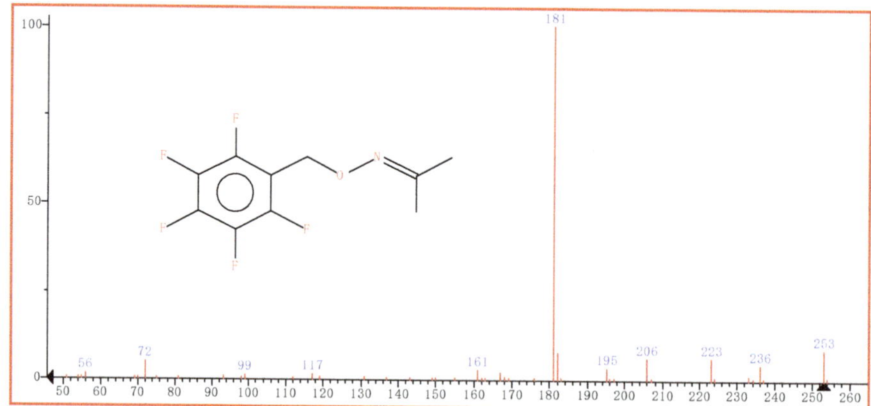

Figure 4. The gas chromatography-mass spectrometry (GC-MS) mass spectrum of acetone-oxime.

Determination of acetone in breath using HS-SPME/GC-MS with on-fiber derivatization yielded satisfactory precision and an excellent sensitivity with a simple procedure. The present method is a potential tool for a non-invasive diagnosis and monitoring of diabetes mellitus and ketoacidosis.

3.3. Correlation Studies of Breath Acetone and Blood Ketone Bodies

We observed the correlation coefficient of $r = 0.897$ between breath acetone and plasma acetoacetate, and we further observed a good correlation of $r = 0.821$ between breath acetone and plasma beta-hydroxybutyrate. This shows a positive indicator of using acetone as a non-invasive biomarker of diabetes mellitus. The blood glucose and acetone also demonstrated a good correlation. The results are found in the Supplementary Material, Figure S2.

4. Discussion

To our knowledge, this study demonstrates for the first time that blood ketone bodies correlate very well with breath acetone in type 2 diabetes mellitus patients. Many studies have reported on either the correlation or analysis of ketone bodies in the blood and breath of type 1 diabetes mellitus, but have never reported ketone bodies on type 2 diabetes mellitus patients. Generally, blood glucose, acetoacetate, beta-hydroxybutyrate and acetone levels differ from individual to individual, and it also varies from day to day, even for the same individual, as can be seen in Figure 3. It depends on the everyday diet, medications, stress and physical activities [31]. In this study, different diabetic and non-diabetic mellitus patients showed the above-mentioned characteristics, thus all the non-diabetic patients showed different plasma metabolites levels. The plasma glucose mean value of 8.55 mmol/L in diabetic and 5.72 mmol/L in non-diabetic patients were observed.

Some diabetic patients showed very good plasma glucose, total cholesterol, triglycerides, HDL cholesterol and LDL cholesterol levels even if they were diagnosed with diabetes mellitus. Thus these patients control their diet, medication and exercise well. Whilst in other diabetic patients, uncontrolled plasma blood glucose levels (11.30, 28.30, 10.30, 10.80, 10.00 mmol/L) were observed. Some patients showed very low plasma glucose levels (3.90 mmol/L) which indicates the state of hypoglycemia that is usually observed in type 1 diabetes mellitus patients. The plasma glucose confirms that the patients are diabetic, and some patients can monitor their glucose level.

Checking for blood glucose alone does not give a clear state of diabetic danger. Hyperketonemia in diabetes is due to insufficient insulin action. It has also been observed in other endocrine-related diseases where excess hormone secretion antagonizes insulin action [14]. Using the Abcam® acetoacetate and beta-hydroxybutyrate assay kits on 30 diabetes and 28 non-diabetes patients, we observed a physiological amount of ketones bodies with the mean values of 0.09 and 0.46, respectively, for diabetes, and 0.05 and 0.44, respectively, for non-diabetes patients. Furthermore, beta-hydroxybutyrate and acetoacetate concentration might provide more information about the severity of ketoacidosis, whether it is related to diabetes, alcohol, or starvation [32]. A blood ketone level less than 0.5 mmol/L is considered to be physiological, whereas hyperketonemia is defined by a value greater than 1 mmol/L, and ketoacidosis is considered to be probable above 3 mmol/L [31,33]. We did not observe hyperketonemia in this study group. However, this does not mean that type 2 diabetes does not undergo hyperketonemia, but simply implies that the patients are able to control their disease. It has been reported that ketone bodies were higher in insulin-dependent patients than non-insulin dependent patients. However, they have found a good correlation of ketone bodies and skin acetone even in controlled diabetes. Our study also found a good correlation in some controlled type 2 diabetes.

Quantifying breath acetone is of importance to this study, as we hope to find the significant correlation between plasma ketone and breathe acetone. Thus, it will strengthen the movement of finding a portable chemoresistive acetone sensor device that will be able to detect acetone from the human breath from as little as 0.1 ppmv. While other non-invasive methods of detection exist such as urine, breath is a less complicated mixture than urine in a sense that it is amenable to complete the analysis of all compounds present. Thus no workup of breath samples are required, in contrast to many analyses performed on urine samples. Additionally, it provides direct information on the respiratory function that is not obtainable by other means. Using HS-SPME/GC-MS, we successfully quantified the acetone level in the breath of both diabetic and non-diabetic mellitus patients. The reaction of acetone in the breath with a derivatizing agent (PFBHA) forms very stable acetone-oxime that was presented on the mass spectrum of the acetone oxime with a base peak at m/z 181. Acetone concentration higher than 1.8 ppmv was found in diabetic breath (Figure 3a). For non-diabetic breath, acetone concentrations lower than 0.8 ppmv were observed (Figure 3b), and the GC-MS spectrum is within the Supplementary results. This study is consistent with the literature [9,14,32–34].

After successfully determining the plasma concentration of acetoacetate, β-hydroxybutyrate and breath acetone, it was necessary to check the correlation of the blood ketones with breath acetone. The diversity of ketone bodies among 30 diabetes patients appeared at baseline (Figure 5). Significant positive correlations between breath acetone and blood AcAc and between breath acetone and blood β-OHB were observed at baseline (R = 0.897 and R = 0.821). This shows a positive indicator of using acetone as a non-invasive biomarker of diabetes mellitus. There are many hypotheses to explain the relationship. One reason being that acetone is a metabolite produced after enzymatic decarboxylation of AcAc, which is in equilibrium with β-OHB via an enzymatic-controlled process by β-OHB dehydrogenase [9]. Although an exponential relationship between acetone and β-OHB, and acetone and AcAc, were observed, acetone reflected overall ketone metabolite concentrations in diabetic patients. This is due to the fact that acetone presents positive deviations from well-known gas/liquid partition laws, such as Henry's law or Raoul's law.

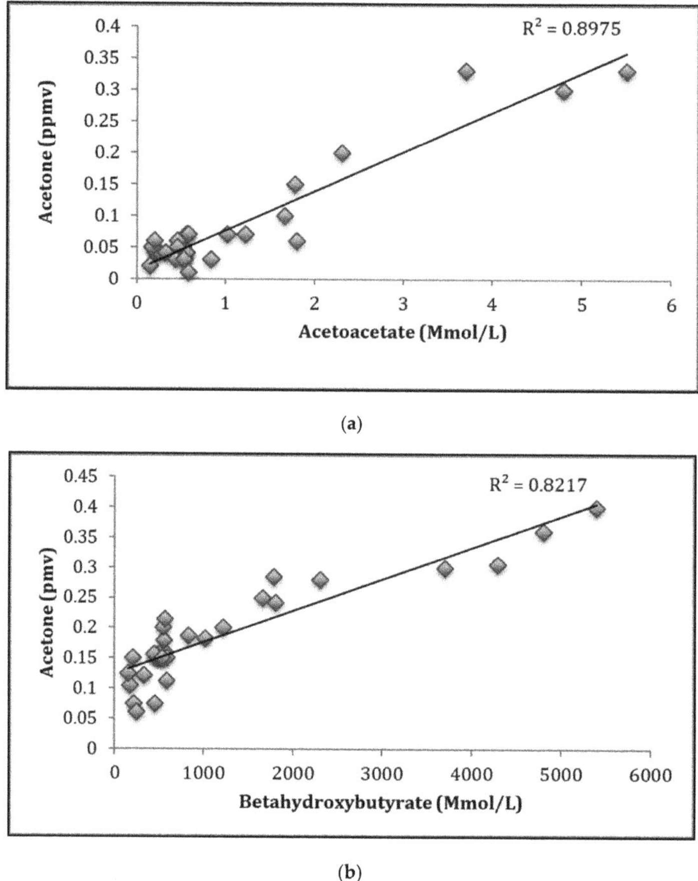

Figure 5. (**a**) Correlation between breath acetone and acetoacetate; (**b**) Correlation between breath acetone and beta-hydroxybutyrates. The correlations were calculated using linear regression.

5. Conclusions

The HS-SPME/GC-MS was used to successfully quantify the amount of breath acetone in type 2 diabetes mellitus and non-diabetes patients. Blood glucose and ketone bodies were also measured. The high amount of ketone bodies (acetone, acetoacetate and beta-hydroxybutyrate) were observed in diabetic patients as opposed to non-diabetic mellitus patients. Breath acetone levels were found to increase with blood β-hydroxybutyrate and blood acetoacetate levels. This might suggest a potential to develop breath gas analysis diagnostic tools to provide an alternative to blood testing for both type 1 and type 2 diabetes mellitus monitoring, and to assist with the prevention of diabetic ketoacidosis.

Supplementary Materials: The following are available online at http://www.mdpi.com/2075-4418/9/4/224/s1, Figure S1: Reconstructed GC-MS ion chromatograms (*m/z* 181) of patient breath samples without insulin injection (a), diabetic breath with insulin (b), and non-diabetic breath (c) sampled using on-fiber SPME derivatization with PFBHA. Figure S2: The measured breath acetone concentration by SPME GC/MS and versus blood glucose in diabetic patients.

Author Contributions: V.S. collected and analyzed the data and prepared the manuscript. M.B., Y.L., and B.M. contributed with reviewing, editing and, supervision with final editing of the manuscript.

Funding: This research was funded by [DSI-CSIR] grant number [CHGER85x].

Acknowledgments: The authors would like to acknowledge Mahomedy Tasneem from Hele Joseph hospital and the patients for volunteering to be part of the study subjects.

Conflicts of Interest: The authors declare no conflict of interest.

References

1. Yamada, K.; Ohishi, K.; Gilbert, A.; Akasaka, M.; Yoshida, N.; Yoshimura, R. Measurement of natural carbon isotopic composition of acetone in human urine. *Anal. Bioanal. Chem.* **2016**, *408*, 1597–1607. [CrossRef] [PubMed]
2. Pereira, J.; Porto-Figueira, P.; Cavaco, C.; Taunk, K.; Rapole, S.; Dhakne, R.; Nagarajaram, H.; Cmara, J.S. Breath analysis as a potential and non-invasive frontier in disease diagnosis: An overview. *Metabolites* **2015**, *5*, 3–55. [CrossRef] [PubMed]
3. Sun, M.; Wang, Z.; Yuan, Y.; Chen, Z.; Zhao, X. Continuous Monitoring of Breath Acetone, Blood Glucose and Blood Ketone in 20 Type 1 Diabetic Outpatients Over 30 Days. *J. Anal. Bioanal. Tech.* **2017**, *8*, 2155–9872. [CrossRef]
4. Newton, C.A.; Raskin, P. Diabetic ketoacidosis in type 1 and type 2 diabetes mellitus: Clinical and biochemical differences. *Arch. Intern. Med.* **2004**, *164*, 1925–1931. [CrossRef]
5. Reyes-Reyes, A.; Horsten, R.C.; Urbach, H.P.; Bhattacharya, N. Study of the exhaled acetone in type 1 diabetes using quantum cascade laser spectroscopy. *Anal. Chem.* **2014**, *87*, 507–512. [CrossRef]
6. Tassopoulos, C.N.; Barnett, D.; Fraser, T.R. Breath-acetone and blood-sugar measurements in diabetes. *Lancet* **1969**, *293*, 1282–1286. [CrossRef]
7. Turner, C.; Walton, C.; Hoashi, S.; Evans, M. Breath acetone concentration decreases with blood glucose concentration in type I diabetes mellitus patients during hypoglycaemic clamps. *J. Breath Res.* **2009**, *3*, 046004. [CrossRef]
8. Anderson, J.C. Measuring breath acetone for monitoring fat loss. *Obesity* **2015**, *23*, 2327–2334. [CrossRef]
9. Prabhakar, A.; Quach, A.; Wang, D.; Zhang, H.; Terrera, M.; Jackemeyer, D.; Xian, X.; Tsow, F.; Tao, N.; Forzanil, E. Breath acetone as biomarker for lipid oxidation and early ketone detection. *Glob. J. Obes. Diabetes Metab. Syndr.* **2014**, *1*, 12.
10. Laffel, L. Ketone bodies: A review of physiology, pathophysiology and application of monitoring to diabetes. *Diabetes Metab. Res.* **1999**, *15*, 412–426. [CrossRef]
11. Henderson, M.J.; Karger, B.; Wrenshall, G. Acetone in the breath. *Diabetes* **1952**, *1*, 188–193. [CrossRef] [PubMed]
12. Wang, C.; Mbi, A.; Shepherd, M. A study on breath acetone in diabetic patients using a cavity ringdown breath analyzer: Exploring correlations of breath acetone with blood glucose and glycohemoglobin A1C. *IEEE Sens. J.* **2010**, *10*, 54–63. [CrossRef]
13. Fan, G.; Yang, C.; Lin, C.; Chen, C.; Shih, C. Applications of Hadamard transform-gas chromatography/mass spectrometry to the detection of acetone in healthy human and diabetes mellitus patient breath. *Talanta* **2014**, *120*, 386–390. [CrossRef] [PubMed]
14. Deng, C.; Zhang, J.; Yu, X.; Zhang, W.; Zhang, X. Determination of acetone in human breath by gas chromatography–mass spectrometry and solid-phase microextraction with on-fiber derivatization. *J. Chromatogr. B* **2004**, *810*, 269–275. [CrossRef]
15. Shaw, J.E.; Sicree, R.A.; Zimmet, P.Z. Global estimates of the prevalence of diabetes for 2010 and 2030. *Diabetes Res. Clin. Pract.* **2010**, *87*, 4–14. [CrossRef]
16. Motala, Y.A.; Fraser, M.A.K.; Pirie, J. Epidemiology of Type 1 and Type 2 Diabetes in Africa. *Eur. J. Prevent. Cardiol.* **2003**, *10*, 77–83. [CrossRef]

17. Bello-Sani, F.; Bakari, A.G.; Anumah, F.E. Dyslipidaemia in persons with type 2 diabetes mellitus in Kaduna, Nigeria. *Int. J. Diabetes Metab.* **2007**, *15*, 9. [CrossRef]
18. Saasa, V.; Malwela, T.; Beukes, M.; Mokgotho, M.; Liu, C.; Mwakikunga, B. Sensing Technologies for Detection of Acetone in Human Breath for Diabetes Diagnosis and Monitoring. *Diagnostics* **2018**, *8*, 12. [CrossRef]
19. Blaikie, T.P.; Edge, J.A.; Hancock, G.; Lunn, D.; Megson, C.; Peverall, R.; Richmond, G.; Ritchie, G.A.; Taylor, D. Comparison of breath gases, including acetone, with blood glucose and blood ketones in children and adolescents with type 1 diabetes. *J. Breath Res.* **2014**, *8*, 046010. [CrossRef]
20. Stephens, J.M.; Sulway, M.J.; Watkins, P.J. Relationship of blood acetoacetate and 3-hydroxybutyrate in diabetes. *Diabetes* **1971**, *20*, 485–489. [CrossRef]
21. Newsholme, P.; Curi, R.; Gordon, S.; Newsholme, E.A. Metabolism of glucose, glutamine, long-chain fatty acids and ketone bodies by murine macrophages. *Biochem. J.* **1986**, *239*, 121. [CrossRef] [PubMed]
22. Minh, T.D.; Oliver, S.R.; Ngo, J.; Flores, R.; Midyett, J.; Meinardi, S.; Carlson, M.K.; Rowland, F.S.; Blake, D.R.; Galassetti, P.R. Noninvasive measurement of plasma glucose from exhaled breath in healthy and type 1 diabetic subjects. *Am. J. Physiol. Endocrinol. Metab.* **2011**, *300*, E1166–E1175. [CrossRef]
23. Novak, B.J.; Blake, D.R.; Meinardi, S.; Rowland, F.S.; Pontello, A.; Cooper, D.M.; Galassetti, P.R. Exhaled methyl nitrate as a noninvasive marker of hyperglycemia in type 1 diabetes. *Proc. Natl. Acad. Sci. USA* **2007**, *104*, 15613–15618. [CrossRef] [PubMed]
24. Jones, A.W.; Sagarduy, A.; Ericsson, E.; Arnqvist, H.J. Concentrations of acetone in venous blood samples from drunk drivers, type-I diabetic outpatients, and healthy blood donors. *J. Anal. Toxicol.* **1993**, *17*, 182–185. [CrossRef]
25. Klocker, A.A.; Phelan, H.; Twigg, S.M.; Craig, M.E. Blood β-hydroxybutyrate vs. urine acetoacetate testing for the prevention and management of ketoacidosis in Type 1 diabetes: A systematic review. *Diabetic Med.* **2013**, *30*, 818–824. [CrossRef]
26. Balasse, E.O.; Féry, F. Ketone body production and disposal: Effects of fasting, diabetes, and exercise. *Diabetes Metab.* **1989**, *5*, 247–270. [CrossRef]
27. Yamane, N.; Tsuda, T.; Nose, K.; Yamamoto, A.; Ishiguro, H.; Kondo, T. Relationship between skin acetone and blood β-hydroxybutyrate concentrations in diabetes. *Clin. Chim. Acta* **2006**, *365*, 325–329. [CrossRef] [PubMed]
28. Taboulet, P.; Deconinck, N.; Thurel, A.; Haas, L.; Manamani, J.; Porcher, R.; Schmit, C.; Fontaine, J.; Gautier, J. Correlation between urine ketones (acetoacetate) and capillary blood ketones (3-beta-hydroxybutyrate) in hyperglycaemic patients. *Diabetes Metab.* **2007**, *33*, 135–139. [CrossRef]
29. Byrne, H.A.; Tieszen, K.L.; Hollis, S.; Dornan, T.L.; New, J.P. Evaluation of an electrochemical sensor for measuring blood ketones. *Diabetes Care* **2000**, *23*, 500–503. [CrossRef]
30. Osuna, E.; Vivero, G.; Conejero, J.; Abenza, J.M.; Martínez, P.; Luna, A.; Pérez-Cárceles, M.D. Postmortem vitreous humor β-hydroxybutyrate: Its utility for the postmortem interpretation of diabetes mellitus. *Forensic Sci. Int.* **2005**, *153*, 189–195. [CrossRef]
31. Kanikarla-Marie, P.; Jain, S.K. Hyperketonemia and ketosis increase the risk of complications in type 1 diabetes. *Free Radic. Biol. Med.* **2016**, *95*, 268–277. [CrossRef] [PubMed]
32. Nasution, T.I.; Nainggolan, I.; Hutagalung, S.D.; Ahmad, K.R.; Ahmad, Z.A. The sensing mechanism and detection of low concentration acetone using chitosan-based sensors. *Sens. Actuat. B Chem.* **2013**, *177*, 522–528. [CrossRef]
33. Wang, L.; Kalyanasundaram, K.; Stanacevic, M.; Gouma, P. Nanosensor device for breath acetone detection. *Sens. Lett.* **2010**, *8*, 709–712. [CrossRef]
34. Xiao, T.; Wang, X.; Zhao, Z.; Li, L.; Zhang, L.; Yao, H.; Wang, J.; Li, Z. Highly sensitive and selective acetone sensor based on C-doped WO3 for potential diagnosis of diabetes mellitus. *Sens. Actuat. B Chem.* **2014**, *199*, 210–219. [CrossRef]

Article

Two Potential Clinical Applications of Origami-Based Paper Devices

Zong-Keng Kuo [1], Tsui-Hsuan Chang [2], Yu-Shin Chen [1], Chao-Min Cheng [2,*] and Chia-Ying Tsai [3,4,*]

1 Institute of Nanoengineering and Microsystems, National Tsing Hua University, Hsinchu 30013, Taiwan; ayudivac@gmail.com (Z.-K.K.); ericys2004@gmail.com (Y.-S.C.)
2 Institute of Biomedical Engineering, National Tsing Hua University, Hsinchu 30013, Taiwan; allmyown74@gmail.com
3 Department of Ophthalmology, Fu Jen Catholic University Hostpital, Fu Jen Catholic University, New Taipei City 24352, Taiwan
4 School of Medicine, College of Medicine, Fu Jen Catholic University, New Taipei City 24205, Taiwan
* Correspondence: chaomin@mx.nthu.edu.tw (C.-M.C.); chiaying131@gmail.com (C.-Y.T.)

Received: 2 November 2019; Accepted: 25 November 2019; Published: 26 November 2019

Abstract: Detecting small amounts of analyte in clinical practice is challenging because of deficiencies in specimen sample availability and unsuitable sampling environments that prevent reliable sampling. Paper-based analytical devices (PADs) have successfully been used to detect ultralow amounts of analyte, and origami-based PADs (O-PADs) offer advantages that may boost the overall potential of PADs in general. In this study, we investigated two potential clinical applications for O-PADs. The first O-PAD we investigated was an origami-based enzyme-linked immunosorbent assay (ELISA) system designed to detect different concentrations of rabbit IgG. This device was designed with four wing structures, each of which acted as a reagent loading zone for pre-loading ELISA reagents, and a central test sample loading zone. Because this device has a low limit of detection (LOD), it may be suitable for detecting IgG levels in tears from patients with a suspected viral infection (such as herpes simplex virus (HSV)). The second O-PAD we investigated was designed to detect paraquat levels to determine potential poisoning. To use this device, we sequentially folded each of two separate reagent zones, one preloaded with NaOH and one preloaded with ascorbic acid (AA), over the central test zone, and added 8 μL of sample that then flowed through each reagent zone and onto the central test zone. The device was then unfolded to read the results on the test zone. The three folded layers of paper provided a moist environment not achievable with conventional paper-based ELISA. Both O-PADs were convenient to use because reagents were preloaded, and results could be observed and analyzed with image analysis software. O-PADs expand the testing capacity of simpler PADs while leveraging their characteristic advantages of convenience, cost, and ease of use, particularly for point-of-care diagnosis.

Keywords: origami-based paper analytic device; origami ELISA; IgG; paraquat

---

1. Introduction

To improve the operation and expand the testing capacity and scope of paper-based analytical devices, we borrowed from the art of origami and folded papers into functional forms that facilitated the application of multiple reagents for conducting more complex and potentially more impactful paper-based, point-of-care biochemical analyses, including multiple, simultaneous chemical reaction-based assays and enzyme-linked immunosorbent assays (ELISAs). In recent years, paper-based analytical devices have demonstrated a variety of advantages for point-of-care diagnostics. Such devices are inexpensive, easily obtained, ecofriendly, naturally wicking, highly compatible with bioassays, and require only small sample amounts [1,2]. Accordingly, a wide array of bioassays have been developed using paper-based analytical devices, including ELISA, and commercialized rapid tests for influenza, bacterial infection, and pregnancy. For example, paper-based ELISA (P-ELISA) devices have been applied to diagnose biological sample protein targets, such as HIV [3], VEGF in aqueous humor [4–6], lactoferrin in tears [7], autoimmune antibodies in serum and blister fluid [8], human chorionic gonadotropin (hCG) in urine samples [9], a cancer marker (prostate-specific antigen, PSA) in serum [10], and *Escherichia coli* in water [11]. The list of analytes identifiable in urine or serum samples using paper-based analytical devices and biochemical analysis includes proteins, glucose, lactate, uric acid, pesticides, and others [4,12,13]. Although paper-based analytical devices are very competitive in terms of cost and sample requirements, they may yet be improved upon. Some P-ELISA procedures, for example, require a number of different reagents and complicated procedures that can be disadvantageous for point-of-care (POC) testing. Further, a more critical environment for biochemical reactions may be required for colorimetric detection of several analytes such as paraquat, a poisonous organophosphate [4,13], and accuracy may suffer when using small sample amounts in paper-based analytic devices. Here, we investigate the possibilities of a user-friendly origami-based paper device to ameliorate P-ELISA complexity and provide a more suitable environment for biochemical analysis.

Keratitis is a leading cause of ocular blindness globally. Gram-positive and gram-negative bacteria, viruses such as herpes simplex virus (HSV), and parasites such as *Acanthamoeba* are known to cause keratitis. Diagnosing keratitis from nonbacterial pathogens, i.e., from HSV or *Acanthamoeba*, is difficult, relies on clinical symptoms and signs, and requires specific treatment such as antiviral agents or chlorhexidine [14,15]. Early and specific diagnosis of keratitis and its cause is important for prognosis [16]. This may be more easily achieved with new methodology. Tears provide the first line of immunological defense for the ocular surface, and tears contain many proteins and cytokines that might be measured as markers of the local immunological state. Secretory immunoglobulins in tears, e.g., IgA, IgG, and IgM, are thought to be specific antimicrobial substances that may increase following ocular surface infection [17]. Secretory IgA, in particular, has been shown to protect the ocular surface from viral and bacterial infection, as well as from parasite infestation, and may operate by coating pathogenic microorganisms to prevent them from adhering to the corneal epithelium [18,19]. IgG and IgM are present in very low concentrations in tears, and IgG concentration is known to increase in inflammation [17–20]. In HSV infection, IgG could sensitize HSV, and lead to increased viral load [21,22] and secretory IgA may neutralize HSV [23]. The effect of neutralization of HSV in ocular tissue may depend on the concentration ratios of IgA and IgG [23]. One study showed similar total serum IgG level but significantly higher anti-*Acanthamoeba* IgG antibody levels in patients with *Acanthamoeba* keratitis compared to those in normal subjects [24]. Patients with *Acanthamoeba* keratitis also displayed lower levels of IgA in tears compared to normal subjects [24]. Efficient methods to detect levels and ratios of each of these compounds would be useful for diagnosis and treatment.

Paraquat is a commonly used herbicide around the world due to its low cost and ready accessibility, especially in developing countries [25]. It has been prohibited in many countries due to its lethal toxicity, but in many developing countries, paraquat is still widely used. Lethally exposed patients often die from multiple organ failure, and without adequate antidote, the mortality rate from exposure is as high as 60–80% [26–28]. In one study, repeat pulse therapy within 5 h of paraquat ingestion decreased the mortality rate to less than 42.9% [29]. To achieve treatment within 5 h, rapid diagnosis of paraquat is invaluable. A paper-based device has been developed to detect paraquat in human serum and could be used in developing countries [6,13]. In the process of detecting paraquat, it is very important to keep the samples wet to ensure accuracy, but this is difficult to do with the existing paper-based device. A folded device that protects the testing zone from drying out may be more useful for maintaining a moist environment that would be more conducive to accurate testing.

Origami is the ancient art of folding flat paper to fabricate three-dimensional sculptures or structures [13,30]. This ancient art form has recently been used with new, scientific intent, as several origami-influenced, paper-based analytical devices (O-PADs) have been developed by folding a single sheet of flat paper into elegantly functional designs using a single patterning step. This method eliminated the need for complicated, sequential, layer-by-layer stacking of individual layers of paper using double-sided tape [31]. In addition, these microfluidic O-PADs can be unfolded to reveal each layer for easy test result analysis [31]. The main purpose for leveraging origami in these studies was to simplify the fabrication of three-dimensional (3D) microfluidic channels or multiple working zones within paper microfluidic devices. Despite some design advances, these devices could not ameliorate the need for multiple reagent preloading or provide the optimal environment for detection [32,33]. Several studies have been undertaken to investigate the possibilities of using origami to fabricate PADs for performing bioassays [31–35], but each did require the use of multiple reagents. In this manuscript we describe our research into possible approaches for optimizing O-PADs that could be useful for continued research efforts.

We explored two particular potential clinical opportunities for POC O-PADS to demonstrate their advantages: (1) The development of an O-PAD ELISA for IgG level testing that reduced test complexity and the number of required reagents; (2) the development of an O-PAD for biochemical analysis that used paraquat as the proof-of-concept assay target. We believe our work could lay some groundwork for the use of O-PADs as an improvement to PADs for POC testing.

2. Materials and Methods

2.1. Design of Origami-Based PAD for ELISA

The O-PAD for ELISA (O-ELISA) was designed on Whatman qualitative filter paper No. 1 and patterned using a wax printer (Xerox Phaser 8650N color printer, Norwalk, CT, USA). Four wings were fashioned to house ELISA reagent loading zones for pre-loading, and the central area was designed and reserved as a testing zone for sample loading. After heating at 105 °C for 3 min, the melted wax wicked through the paper to create hydrophobic barrier wells in the paper. Reagents were then pre-loaded into reagent zone barrier wells. These reagents included the following: (1) 2 µL of blocking buffer (5% (*w*/*v*) bovine serum albumin (BSA) (Sigma, SI-A7906, St. Louis, MO, USA) in PBS (Corning, 21-040-CM, Corning, NY, USA); (2) 2 µL solution of alkaline phosphatase (ALP)-conjugated detection antibody (Cell signaling, #7054, Danvers, MA, USA) (20 µg/mL, 0.01% (*v*/*v*) Tween-20 (Sigma, P9416, St. Louis, MO, USA); and (3) 2 µL BCIP/NBT substrate (13.4 mM BCIP (Sigma, B6274), 9 mM NBT (Sigma, N5514), 25 mM $MgCl_2$ (Sigma, M8266), 500 mM NaCl (Sigma, S7653) in 500 mM Tris buffer pH 9.5 (Sigma, T4661). After loading each reagent, the device was placed under ambient conditions for 5 to 10 min until the reagents dried.

2.2. Performing Origami-Based PAD for ELISA

The process for creating and using O-PADS is outlined in Figure 2. Briefly, we first loaded 2 µL of sample into the testing zone and allowed it to dry for 10 min under ambient conditions. Then, we folded the reagent zone to contact the testing zone, and loaded 3 µL of PBS into the reagent zone and allowed PBS to transfer reagents to the testing zone. The reaction time for BSA blocking was 5 min and the time for reaction with ALP-conjugated antibody was 10 min. Then, we folded the wash zone to the testing zone, placed the O-PAD on paper towels and washed with 3 µL of PBS. Finally, we folded the substrate zone to the testing zone and loaded 3 µL of PBS to transfer substrate to the testing zone and allowed the enzymatic reaction to proceed for 20 min under ambient conditions. The O-PAD was subsequently scanned using a photo scanner (EPSON 3490 PHOTO). Gray-scale color intensity following the enzymatic reaction was quantified using ImageJ software (ImageJ 1.80, National Institutes of Health, Bethesda, MD, USA). In order to diminish both background noise and analytical variation, we subtracted the original background value of each test zone for each detection result. The experimental data were further analyzed using Sigmaplot (version 13.0, Systat Software, Inc., San Jose, CA, USA).

2.3. Fabrication of the Origami-Based PAD for Paraquat Detection

An O-PAD for paraquat detection is shown in Figure 3. The pattern created on the device was fabricated as described above. We preloaded 5 µL of 5N NaOH and 6 µL of 5% (*w*/*v*) ascorbic acid onto the left and right parts of the PAD, respectively. After loading each reagent, the device was placed under ambient conditions for 20 min until the reagents dried.

2.4. Performing Origami-Based Detection of Paraquat

To perform origami-based detection of paraquat, we sequentially folded the left and right parts of the structure over and onto the central portion. We then loaded 8 µL of sample onto the test zone and allowed it to rest for 10 min at 25 °C. Detection results were recorded using a digital camera and the signal was analyzed using ImageJ software. The RGB color value of each detection result was obtained, and the R value was used for further calculation [13,36]. We subtracted original test zone background values for each detection result.

3. Results and Discussion

3.1. O-PAD for ELISA

We designed and fabricated a P-ELISA device with multiple preloaded reagents and, borrowing from origami, folded it to make a multilayered, three-dimensional device that provided structural advantages. The design of this O-PAD device is provided in Figure 1a,b.

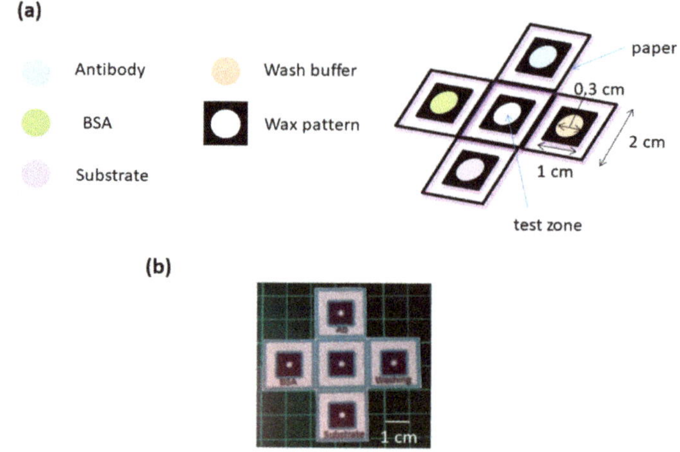

Figure 1. The design of an origami-influenced paper-based analytic device (O-PAD) for enzyme-linked immunosorbent assay (ELISA). (**a**) Reagents were pre-loaded into the four-wing fashioned zones, and the central area was reserved as testing zone for sample loading. (**b**) The picture of our O-PAD.

We used rabbit IgG as the analyte for demonstrating device performance (see schematic illustration in Figure 2. We used serial dilutions of IgG diluted in PBS (260 nM–2.6 pM) to investigate the limit of detection (LOD). The total time to complete testing with this novel device was approximately 46 min. Results from our test were scanned using a photo scanner and the mean gray-scale intensity of the color formed was measured using ImageJ (Figure 3). The LOD for this rabbit IgG system was calculated using the generated colorimetric results. We graphed colorimetric intensity versus rabbit IgG concentration (log scale) to produce the standard curve of this rabbit IgG system. We further employed nonlinear regression using the Hill equation, to generate a sigmoidal curve fit that has previously been used to describe the antibody–antigen reaction kinetics in paper [3]. Similar sigmoidal fits for both rabbit IgG [3] and human vascular endothelial growth factor [6] paper-based ELISA studies have been demonstrated. In the Hill equation (Equation (1)), è represents the fraction of occupied binding sites, [L] represents the ligand concentration (in mol), $[L_{50}]$ describes the ligand concentration when half of the binding sites are occupied (in mol), and n is the Hill coefficient. The fraction of occupied binding sites can be described by the ratio of the observed intensity (I) of the colorimetrical signal to the maximum intensity (Equation (2)). The intensity is proportional to the amount of detected antigens (Equation (3)).

$$\grave{e} = [L]^n/([L]^n + [L_{50}]^n) \quad (1)$$

$$\grave{e} = I/I_{max} \quad (2)$$

$$I = I_{max}[L]^n/([L]^n + [L_{50}]^n) \quad (3)$$

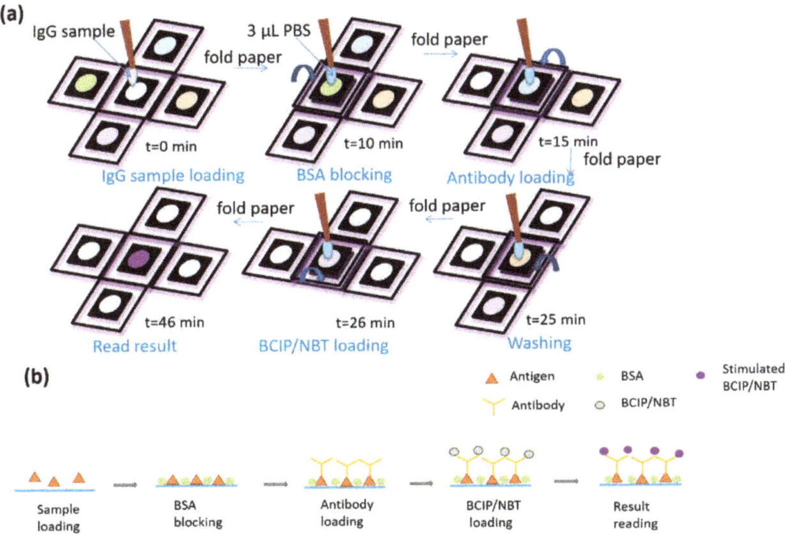

Figure 2. Schematic illustrations demonstrating use of O-PAD ELISA. (**a**) Reagents were pre-loaded into reagent zone barrier wells. These reagents included the following: (1) 2 μL of BSA blocking buffer (2) 2 μL solution of alkaline phosphatase (ALP)-conjugated detection antibody (3) 2 μL BCIP/NBT substrate. After loading each reagent, the device was placed under ambient conditions for 5 to 10 min until the reagents dried. (**b**) Schematic illustrations for BSA blocking, antibody loading, and BCIP/NBT loading.

Accordingly, we fitted the generated colorimetric results into the standard curve using Sigmaplot, and determined that the LOD of the IgG system using our device was about 201 pM, which was calculated by fitting the signal (25.4) that was three times the standard deviation of the blank control [3]. This low IgG LOD could be beneficial for analysis of low-volume specimens in clinical practice. Tear fluid for instance, an example of a low-volume sampling source, demonstrates increased inflammation caused by giant papillary conjunctivitis with an IgG concentration of 160 nM (i.e., IgG detection limit of 160 nM) [37] and viral infection, such as HSV [38], and could benefit from the availability of such a potential diagnostic tool.

Origami has previously been applied to fabricate a paper-based device for performing bioassays [32,33,39]. While the process requires the use of multiple reagents and complex steps, end-user time is saved by preloading with the necessary reagents for performing ELISA. With this ELISA O-PAD, users only need to load their sample to the test zone and apply a single, fixed volume of buffer to initiate testing. Results can be easily obtained via colorimetric readout. Moreover, the colorimetric signal can be scanned or photographed using a smartphone and the results more finely analyzed and compared. The mean color formation intensity can be quantified using versatile software applications including ImageJ or APPs, which can provide convenient POC diagnostic results.

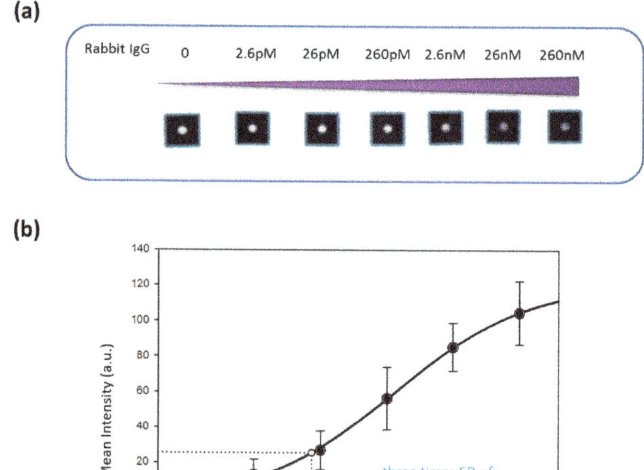

Figure 3. Results of rabbit IgG detection using the O-PAD. (**a**) Colorimetric response. (**b**) Calibration curve of rabbit IgG using mean gray-scale intensity versus several concentrations of rabbit IgG ($n = 6$).

Although an O-PAD with pre-loaded reagents is convenient for POC testing, the storage requirements of proteins, enzymes, substrates, and reagents may limit practicality. We used alkaline phosphatase-conjugated antibody as the detection antibody because commercialized alkaline phosphatase was provided to us as lyophilized powder shipped at ambient, room temperature. BCIP/NBT, the colorimetric substrate of alkaline phosphatase, was available in tablet form and could be crushed to a powder. For both of the above reasons, we felt that a preloaded alkaline phosphatase/BCIP/NBT system was suitable for our O-PAD. The horseradish peroxidase (HRP) and 3,3′,5,5′-tetramentylbenzidine (TMB) system commonly used for ELISA kits may be unsuitable for O-PAD development because the hydrogen peroxide involved in the colorimetric reaction is unstable [40], but we noted that Ramachandran et al. reported a method for long-term dry storage of HRP-conjugated antibody as well as its colorimetric substrate, diaminobenzidine (DAB) [41]. Indeed, some enzymes or proteins that have not undergone optimized drying procedures in suitable buffers may not be stable in a dried state for long-term storage. Our manuscript describes a process to leverage the known advantages of P-ELISA in a novel manner by folding it, origami-fashion, and pre-loading multiple reagents to create a multiplexed diagnostic tool. We believe that the versatility of this approach could lead to an exceptionally useful commercial product, especially if long-term dry-state storage solutions can be found for a broad array of test reagents.

A three-dimensional (3D) microfluidic paper-based analytical device (3D-μPAD) was reported by Liu et al. that used a sliding movable test strip that was dipped into reagent stored within the device [37]. After sliding the test strip to a different location under spots preloaded with stored reagent, buffer was loaded and the integrated, stored reagent was transferred to the test zone. ELISA could be completed by repeating this test-strip sliding action and loading additional buffer [37]. Using rabbit IgG as a model analyte, Liu et al. performed an ELISA in 43 min, with a detection limit of 330 pM. Although this device provides an alternative solution to previous P-ELISA methods, it was difficult to control the contact between test zones and stored, preloaded reagents. Moreover, mass fabrication of this particular 3D-μPAD was problematic.

3.2. O-PAD for Paraquat Detection

Our previous results showed that a folded, origami-inspired structure could facilitate efficient PAD procedures. However, successful detection of multiple analytes using a single PAD relied on maintaining particular, sometimes critical, conditions such as moisture level. In a previous study, we showed that paraquat detection in a PAD required a wet environment [13,36]. Traditional ELISA, however, requires application of multiple reagents, and the time to complete such a process using paper increases the likelihood of any moisture-sensitive components drying out. To remedy this we created an O-PAD for paraquat detection that used its protective folds to maintain moisture during the sample testing phase. This O-PAD for paraquat detection, with a printed wax pattern created as described above, is shown in Figure 4. We preloaded 5 μL of 5 N NaOH and 6 μL of 5% (*w*/*v*) ascorbic acid (AA) onto the left and right parts of our PAD, respectively. After preloading each reagent, we placed the device under ambient conditions for 20 min to allow the reagents to dry. We then folded the reagent zones containing NaOH and AA onto the test zone sequentially and loaded 8 μL of test sample onto the test zone before letting the device sit for 10 min at 25 °C. Paraquat, in contact with NaOH and AA, changed to paraquat radical ion (blue color) [13,36]. The resulting colorimetric response for each reagent was recorded using a digital camera and analyzed using ImageJ software. The RGB color value of each detection result was obtained. Because previous studies indicated that R value provided the best colorimetric reaction for evaluative and diagnostic purposes [13,36], we employed the same protocol here. In order to diminish background noise and analytical variation, we subtracted any original background values for each test zone result. We used five different concentrations of paraquat standard solution (i.e., 0, 10, 25, 50, and 100 ppm) to establish our standard curve (Figure 5). The LOD for paraquat detection using our O-PAD was 5.82 ppm, which was calculated by using the signal results of our 0 ppm groups plus three-fold standard deviations in the regression model of the standard curve (Figure 5).

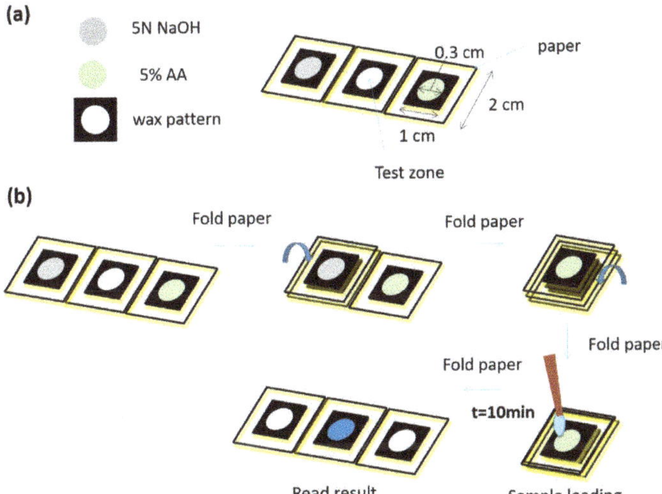

Figure 4. Schematic illustrations of O-PAD for paraquat detection. (**a**) 5 μL of 5 N NaOH and 6 μL of 5% (*w*/*v*) ascorbic acid (AA) were preloaded onto the left and right parts of our PAD, respectively. (**b**) The reagent zones containing NaOH and AA were folded onto the test zone sequentially. 8 μL of test sample was loaded onto the test zone, and the device stood for 10 min at 25 °C. Paraquat, in contact with NaOH and AA, changed to paraquat radical ion (blue color).

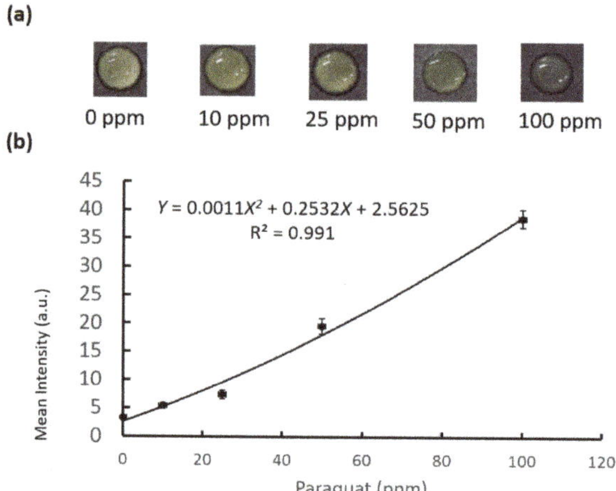

Figure 5. Result of paraquat detection using the O-PAD. (**a**) Colorimetric response. (**b**) The calibration curve of paraquat using mean RGB intensity versus several concentrations of paraquat (n = 3).

The LOD (5.82 ppm) using our O-PAD was comparable to the LOD (3.86 ppm) of a previous 96-well format PAD we developed [13]. The O-PAD was more convenient for users because the relevant reagents were preloaded and the results could be easily obtained following a single test sample loading step. Further, the O-PAD comprising three folded paper layers provided a moist environment for paraquat testing that was more suitable than the 96-well format testing device, which consisted of a single layer of paper. These advantages could provide useful point-of-care device development strategies for other analyte detection models. It is worth noting that the lack of long-term storage stability for preloaded reagents may limit the use of our O-PAD for paraquat detection, i.e., ascorbic acid oxidizes in the presence of oxygen. We believe that methods aimed at improving long-term, dry-state reagent storage could help to optimize and broaden the range of potential applications for O-PADs.

Recently, PADs have demonstrated advantages for POC diagnostics. Paper-based ELISA, for instance, require less sample volume and less time to complete than conventional ELISA because of the high surface-to-volume ratio of paper fibers within the test zones [3]. However, PADs are still plagued by some shortcomings that remain to be resolved (Table 1). For example, multiple reagents and more consumables are required to conduct assays, which increase process complexity. Moreover, single-layer PADs may suffer from dry conditions that are less ideal, even insufficient, for testing particular analytes. To ameliorate these disadvantages, we sought to leverage the folded structure of origami to fabricate user-friendly, sensitive, and accurate PADs. Reagents required for conducting assays on such a device could be preloaded and integrated among the device's protective folds. Use of one of these devices requires relatively straightforward folding and unfolding manipulations. Together, these advantages could increase ease, convenience, and acceptability of such devices especially for POC applications in resource-limited areas. We believe that reliance on an ancient art form, origami, could expand the future potential for the development of PADs for clinical diagnostics that could reap real benefits for human health.

Table 1. Comparisons of conventional ELISA and paper-based ELISA (P-ELISA) to origami-based paper devices (some information adapted from reference [3]).

	Conventional	P-ELISA	Origami
Pre-loaded reagents for assay	No	No	Yes
Reagents needed for conducting assay	Multiple reagents	Multiple reagents	One reagent or sample only
Equipment	Multiple tips	Multiple tips	Single tip
Convenience	Low	Low	High
Environment for reactions of detection	Less stable	Less stable	More stable
Required antigen volume	70 μL	3 μL	2 μL
Total complete time	213 min	51 min	46 min
LOD	5.7 pM	18 nM	201 pM

LOD—limit of detection.

4. Conclusions

O-PADs provide a more convenient, reliable, and easy-to-use approach for target sample detection. In this study, we successfully used an O-PAD to detect rabbit IgG level with a low LOD. Due to the limited information regarding the IgG LOD for ophthalmic viral infection or inflammation, we could not compare our results to the references. Overall, this study demonstrated two potential areas for clinical application of OPADs that could provide significant scientific and health benefits. Our device provides two advantages: (1) to reduce test complexity and the number of required reagents through pre-loaded reagents; and, (2) to provide a wet environment for moisture-sensitive analysis, e.g., paraquat detection. However, there are still several limitations associated with our device. First, while providing advantages over paper-based ELISA, our device still could not reach the sensitivity of conventional ELISA; this limitation may be improved by optimizing the assay to suppress the background signal. Second, the long-term storage stability of preloaded reagents such as ascorbic acid, which oxidizes if exposed to oxygen, may limit usage. We believe that additional developments, especially those aimed at improving long-term dry-state storage, would improve our device and broaden its applicability.

Author Contributions: Conceptualization, Z.-K.K. and C.-M.C.; methodology, C.-M.C.; software, T.-H.C.; validation, Y.-S.C.; formal analysis, T.-H.C.; investigation, Z.-K.K.; resources, C.-M.C.; data curation, Y.-S.C.; writing—original draft preparation, Z.-K.K., C.-Y.T.; writing—review and editing, C.-Y.T., C.-M.C.; visualization, Z.-K.K.; supervision, C.-M.C.; project administration, C.-M.C.; funding acquisition, C.-M.C., C.-Y.T.

Funding: This project is funded by the Ministry and Science Technology, Taiwan, no.: MOST 108-2622-E-007-020 -CC3 and the Fu Jen Catholic University, no: PL201908002B.

Acknowledgments: This study was supported by grants (107-2628-E-007-001-MY3 and 108-2622-E-007-020-CC3) from the Ministry of Science and Technology, Taiwan, and a grant (PL-201809004-V) from the Fu Jen Catholic University Hospital, Taiwan.

Conflicts of Interest: The authors declare no conflict of interest. The funders had no role in the design of the study; in the collection, analyses, or interpretation of data; in the writing of the manuscript, or in the decision to publish the results.

References

1. Chen, Y.H.; Kuo, Z.K.; Cheng, C.M. Paper—A potential platform in pharmaceutical development. *Trends Biotechnol.* **2015**, *33*, 4–9. [CrossRef]
2. Sher, M.; Zhuang, R.; Demirci, U.; Asghar, W. Paper-based analytical devices for clinical diagnosis: Recent advances in the fabrication techniques and sensing mechanisms. *Expert Rev. Mol. Diagn.* **2017**, *17*, 351–366. [CrossRef]
3. Cheng, C.M.; Martinez, A.W.; Gong, J.; Mace, C.R.; Phillips, S.T.; Carrilho, E.; Mirica, K.A.; Whitesides, G.M. Paper-based ELISA. *Angew. Chem. Int. Ed. Engl.* **2010**, *49*, 4771–4774. [CrossRef] [PubMed]
4. Yen, T.H.; Chen, K.H.; Hsu, M.Y.; Fan, S.T.; Huang, Y.F.; Chang, C.L.; Wang, Y.P.; Cheng, C.M. Evaluating organophosphate poisoning in human serum with paper. *Talanta* **2015**, *144*, 189–195. [CrossRef] [PubMed]

5. Hsu, M.Y.; Hung, Y.C.; Hwang, D.K.; Lin, S.C.; Lin, K.H.; Wang, C.Y.; Choi, H.Y.; Wang, Y.P.; Cheng, C.M. Detection of aqueous VEGF concentrations before and after intravitreal injection of anti-VEGF antibody using low-volume sampling paper-based ELISA. *Sci. Rep.* **2016**, *6*, 34631. [CrossRef] [PubMed]
6. Hsu, M.Y.; Yang, C.Y.; Hsu, W.H.; Lin, K.H.; Wang, C.Y.; Shen, Y.C.; Chen, S.F.; Chau, S.F.; Tsai, H.Y.; Cheng, C.M. Monitoring the VEGF level in aqueous humor of patients with ophthalmologically relevant diseases via ultrahigh sensitive paper-based ELISA. *Biomaterials* **2014**, *35*, 3729–3735. [CrossRef]
7. Yamada, K.; Takaki, S.; Komuro, N.; Suzuki, K.; Citterio, D. An antibody-free microfluidic paper-based analytical device for the determination of tear fluid lactoferrin by fluorescence sensitization of Tb3+. *Analyst* **2014**, *139*, 1637–1643. [CrossRef]
8. Hsu, C.K.; Huang, H.Y.; Chen, W.R.; Nishie, W.; Ujiie, H.; Natzuga, K.; Fan, S.T.; Wang, H.K.; Lee, J.Y.Y.; Tsai, W.L.; et al. Paper-based ELISA for the detection of autoimmune antibodies in body fluid-the case of bullous pemphigoid. *Anal. Chem.* **2014**, *86*, 4605–4610. [CrossRef]
9. Apilux, A.; Ukita, Y.; Chikae, M.; Chailapakul, O.; Takamura, Y. Development of automated paper-based devices for sequential multistep sandwich enzyme-linked immunosorbent assays using inkjet printing. *Lab Chip* **2013**, *13*, 126–135. [CrossRef]
10. Wang, S.; Ge, L.; Song, X.; Yu, J.; Ge, S.; Huang, J.; Zeng, F. Paper-based chemiluminescence ELISA: Lab-on-paper based on chitosan modified paper device and wax-screen-printing. *Biosens. Bioelectron.* **2012**, *31*, 212–218. [CrossRef]
11. Shih, C.M.; Chang, C.L.; Hsu, M.Y.; Lin, J.Y.; Kuan, C.M.; Wang, H.K.; Huang, C.T.; Chung, M.C.; Huang, K.C.; Hsu, C.E. Paper-based ELISA to rapidly detect *Escherichia coli*. *Talanta* **2015**, *145*, 2–5. [CrossRef] [PubMed]
12. Dungchai, W.; Chailapakul, O.; Henry, C.S. Use of multiple colorimetric indicators for paper-based microfluidic devices. *Anal. Chim Acta* **2010**, *674*, 227–233. [CrossRef] [PubMed]
13. Kuan, C.M.; Lin, S.T.; Yen, T.H.; Wang, Y.L.; Cheng, C.M. Paper-based diagnostic devices for clinical paraquat poisoning diagnosis. *Biomicrofluidics* **2016**, *10*, 034118. [CrossRef] [PubMed]
14. Barron, B.A.; Gee, L.; Hauck, W.W.; Kurinij, N.; Dawson, C.R.; Jones, D.B.; Wilhelmus, K.R.; Kaufman, H.E.; Sugar, J.; Hyndiuk, R.A. Herpetic Eye Disease Study: A controlled trial of oral acyclovir for herpes simplex stromal keratitis. *Ophthalmology* **1994**, *101*, 1871–1882. [CrossRef]
15. Schwab, I.R.J.O. Oral acyclovir in the management of herpes simplex ocular infections. *Ophthalmology* **1988**, *95*, 423–430. [CrossRef]
16. Bacon, A.S.; Dart, J.K.; Ficker, L.A.; Matheson, M.M.; Wright, P.J.O. Acanthamoeba keratitis: The value of early diagnosis. *Virology* **1993**, *100*, 1238–1243.
17. German, A.; Hall, E.; Day, M.J. Measurement of IgG, IgM and IgA concentrations in canine serum, saliva, tears and bile. *Vet. Immunol.* **1998**, *64*, 107–121. [CrossRef]
18. Bron, A.; Seal, D.J. The defences of the ocular surface. *Trans. Ophthalmol. Soc. UK* **1986**, *105*, 18–25.
19. McClellan, B.H.; Whitney, C.R.; Newman, L.P.; Allansmith, M.R. Immunoglobulins in tears. *Am. J. Ophthalmol.* **1973**, *76*, 89–101. [CrossRef]
20. Coyle, P.; Sibony, P.J. Tear immunoglobulins measured by ELISA. *Investig. Ophth. Vis. Sci.* **1986**, *27*, 622–625.
21. Ashe, W.K.; Mage, M.; Notkins, A.L. Kinetics of neutralization of sensitized herpes simplex virus with antibody fragments. *Short Commun.* **1969**, *37*, 290–293. [CrossRef]
22. Ashe, W.K.; Notkins, A.L. Kinetics of sensitization of herpes simplex virus and its relationship to the reduction in the neutralization rate constant. *Virology* **1967**, *33*, 613–617. [CrossRef]
23. Centifanto, Y.; Kaufman, H.E. Secretory immunoglobulin A and herpes keratitis. *Infect. Immun.* **1970**, *2*, 778–782.
24. Alizadeh, H.; Apte, S.; El-Agha, M.-S.H.; Li, L.; Hurt, M.; Howard, K.; Cavanagh, H.D.; McCulley, J.P.; Niederkorn, J.Y. Tear IgA and serum IgG antibodies against Acanthamoeba in patients with Acanthamoeba keratitis. *Cornea* **2001**, *20*, 622–627. [CrossRef]
25. Wesseling, C.; De Joode, B.V.W.; Ruepert, C.; León, C.; Monge, P.; Hermosillo, H.; Partanen, L.J. Paraquat in developing countries. *Int. J. Occup. Environ. Health* **2001**, *7*, 275–286. [CrossRef]
26. Pond, S.M.; Rivory, L.P.; Hampson, E.C.; Roberts, M.S. Kinetics of toxic doses of paraquat and the effects of hemoperfusion in the dog. *J. Toxicol. Clin. Toxicol.* **1993**, *31*, 229–246. [CrossRef]
27. Sittipunt, C. Paraquat poisoning. *Respir. Care.* **2005**, *50*, 383–385.
28. Weng, C.H.; Hu, C.C.; Lin, J.L.; Lin-Tan, D.T.; Hsu, C.W.; Yen, T.H. Predictors of acute respiratory distress syndrome in patients with paraquat intoxication. *PLoS ONE* **2013**, *8*, e82695. [CrossRef]

29. Hsu, C.W.; Lin, J.L.; Lin.Tan, D.T.; Chen, K.H.; Yen, T.H.; Wu, M.S.; Lin, S.C. Early hemoperfusion may improve survival of severely paraquat-poisoned patients. *PLoS ONE* **2012**, *7*, e48397. [CrossRef]
30. Johnson, M.; Chen, Y.; Hovet, S.; Xu, S.; Wood, B.; Ren, H.; Tokuda, J.; Tse, Z.T.H. Fabricating biomedical origami: A state-of-the-art review. *Int. J. Comput. Assist. Radiol. Surg.* **2017**, *12*, 2023–2032. [CrossRef]
31. Liu, H.; Crooks, R.M. Three-dimensional paper microfluidic devices assembled using the principles of origami. *J. Am. Chem Soc.* **2011**, *133*, 17564–17566. [CrossRef] [PubMed]
32. Ge, L.; Wang, S.; Song, X.; Ge, S.; Yu, J. 3D origami-based multifunction-integrated immunodevice: Low-cost and multiplexed sandwich chemiluminescence immunoassay on microfluidic paper-based analytical device. *Lab Chip* **2012**, *12*, 3150–3158. [CrossRef] [PubMed]
33. Govindarajan, A.V.; Ramachandran, S.; Vigil, G.D.; Yager, P.; Bohringer, K.F. A low cost point-of-care viscous sample preparation device for molecular diagnosis in the developing world; an example of microfluidic origami. *Lab Chip* **2012**, *12*, 174–181. [CrossRef] [PubMed]
34. Chen, C.-A.; Yeh, W.-S.; Tsai, T.-T.; Chen, C.-F. Three-dimensional origami paper-based device for portable immunoassay applications. *Lab Chip* **2019**, *19*, 598–607. [CrossRef] [PubMed]
35. Tian, T.; An, Y.; Wu, Y.; Song, Y.; Zhu, Z.; Yang, C. Integrated Distance-Based Origami Paper Analytical Device for One-Step Visualized Analysis. *ACS Appl. Mater. Interfaces* **2017**, *9*, 30480–30487. [CrossRef]
36. Chang, T.-H.; Tung, K.-H.; Gu, P.-W.; Yen, T.-H.; Cheng, C.-M. Rapid Simultaneous Determination of Paraquat and Creatinine in Human Serum Using a Piece of Paper. *Micromachines* **2018**, *9*, 586. [CrossRef]
37. Donshik, P.C.; Ballow, M. Tear immunoglobulins in giant papillary conjunctivitis induced by contact lenses. *Am. J. Ophthalmol.* **1983**, *96*, 460–466. [CrossRef]
38. Borderie, V.M.; Gineys, R.; Goldschmidt, P.; Batellier, L.; Laroche, L.; Chaumeil, C.J.C. Association of Anti–Herpes Simplex Virus IgG in Tears and Serum With Clinical Presentation in Patients With Presumed Herpetic Simplex Keratitis. *Cornea* **2012**, *31*, 1251–1256. [CrossRef]
39. Liu, X.; Cheng, C.; Martinez, A.; Mirica, K.; Li, X.; Phillips, S.; Mascarenas, M.; Whitesides, G. A portable microfluidic paper-based device for ELISA. In Proceedings of the 2011 IEEE 24th International Conference on Micro Electro. Mechanical Systems, Cancun, Mexico, 23–27 January 2011; pp. 75–78.
40. Verma, M.S.; Tsaloglou, M.-N.; Sisley, T.; Christodouleas, D.; Chen, A.; Milette, J.; Whitesides, G.M. Sliding-strip microfluidic device enables ELISA on paper. *Biosens Bioelectron.* **2018**, *99*, 77–84. [CrossRef]
41. Ramachandran, S.; Fu, E.; Lutz, B.; Yager, P.J.A. Long-term dry storage of an enzyme-based reagent system for ELISA in point-of-care devices. *Analyst* **2014**, *139*, 1456–1462. [CrossRef]

 *diagnostics*

Article

Comparison of Procalcitonin Assays on KRYPTOR and LIAISON® XL Analyzers

Mariella Dipalo *, Cecilia Gnocchi, Paola Avanzini, Roberta Musa, Martina Di Pietro and Rosalia Aloe

SSD Biochimica ad Elevata Automazione, Azienda Ospedaliera Universitaria di Parma, Via Gramsci 14, 43126 Parma, Italy

* Correspondence: mdipalo@ao.pr.it

Received: 5 July 2019; Accepted: 6 August 2019; Published: 8 August 2019

Abstract: Our laboratory performs procalcitonin (PCT) assays on a Brahms KRYPTOR analyzer with the Brahms PCT sensitive Kryptor kit. In this study, we wanted to compare the assays obtained in this way with the ones performed on the LIAISON® XL. From January to May 2017, 171 samples were analyzed, of which 65 from female patients (age: 22–98 years) and 106 from male patients (age: 16–97 years). The PCT determination was performed using the LIAISON® XL and KRYPTOR analyzers, by chemiluminescence (Chemiluminescence immunoassay—CLIA) (LIAISON® BRAHMS PCT® II GEN) and immunofluorescence (Brahms PCT sensitive Kryptor) assay, respectively. For the LIAISON® BRAHMS PCT® II GEN, 52% of the results were placed between 0.0 and 0.5 ng/mL, 18% between 0.5 and 2.0 ng/mL, and 30% between 2.0 and 100 ng/mL; the mean was 4.09 ng/mL, the median 0.456 ng/mL, the maximum value 97.2 ng/mL, and the minimum value 0.02 ng/mL. For the Brahms PCT sensitive Kryptor, 55% of the results were positioned between 0.0 and 0.5 ng/mL, 21% between 0.5 and 2.0 ng/mL, and 24% between 2.0 and 100 ng/mL; the mean was 3.72 ng/mL, the median 0.39 ng/mL, the maximum value 103 ng/mL, and the minimum value 0.01 ng/mL. The mean of the results obtained with the two methods showed no significant differences (3.717 for Kryptor and 4.094 for LIAISON®). PCT assay with Brahms reagents, both on the Kryptor and LIAISON®XL platforms, offers excellent performance in terms of sensitivity and specificity.

Keywords: sepsis; PCT; procalcitonin; immunoassay; antibiotic; chemiluminescence; immunofluorescence

1. Introduction

Sepsis and its consequences are an important cause of mortality and morbidity, both for inpatients and outpatients, in all age groups. Despite the progress made by medicine, diagnosing sepsis today still constitutes a challenge. This is mainly due to the wide variability of the clinical presentation (which, in turn, depends on the site of the initial infectious outbreak), the involved microorganism, the timing of evaluation and clinical intervention, and, finally, the presence of comorbidities and concomitant therapies [1]. Until a few years ago, the C-reactive protein (PCR) assay was the only one that could help with the monitoring of the progress of inflammation, even if in a non-specific way. Blood culture represents the gold standard test for the correct identification of pathogenic species responsible for sepsis, but its execution requires a timeframe that ranges from 24 to 72 h, depending on the complexity of the investigations needed. In recent years, the requests for the determination of procalcitonin (PCT) have been rapidly increasing, also because of the availability of this assay on various analytical platforms using different technologies [2].

Procalcitonin (PCT) is a peptide formed by 116 amino acids (aa), normally synthesized in the C cells of the thyroid, from pre-procalcitonin (composed of 141 aa), which is then converted, once in the blood stream, into a mature form, called calcitonin (composed of 32 aa), involved in the homeostasis of calcium. In patients with severe bacterial infections, PCT synthesis can also occur in other districts (especially in the liver, kidneys, lungs, pancreas, intestine, and leukocytes) due to specific inflammatory stimuli, mainly mediated by interleukin 6 (IL-6) and tumor necrosis factor alpha, which are triggered by the lipopolysaccharide, the major component of the external membrane of Gram-negative bacteria. Under such conditions, the PCT concentration can increase up to 10,000 times compared to the usually normal values, which are very low at below 0.05 ng/mL [3]. This peculiar biological behavior can be used to diagnose serious infections, especially sepsis [4]. The number of studies and meta-analyses evaluating the diagnostic performance of PCT for both the diagnosis and management of sepsis has increased exponentially over the past decade. According to a recent meta-analysis published by Tan et al. [5], PCT shows a diagnostic accuracy of 85% for sepsis (with 0.80 sensitivity and 0.77 specificity), thus presenting a significantly higher precision than the one of C-reactive protein, which stands instead at 73% (with 0.80 sensitivity and 0.61 specificity). More importantly, in another meta-analysis published by Meier et al. [6], the management of antibiotic therapy by means of PCT determination was significantly effective in shortening the duration of the therapy itself (mean change, −2.86 days), thus representing a valuable step towards reducing antibiotic resistance [7]. In general, patients for whom PCT monitoring is required are as follows:

- patients undergoing antibiotic therapy, in order to check the effectiveness of the treatment;
- patients at high risk of severe infections due to the presence of concomitant diseases,
- patients subject to long-term ventilation;
- patients at risk of catheter-related infections;
- patients subject to immunosuppression (oncology, transplant, chemotherapy, neutropenia);
- post-operative or post-traumatic patients;
- patients at high risk of super infections (burns).

The implementation of this test could therefore be of help not only for the early diagnosis of sepsis, but also for the reduction of the risk of antibiotic resistance through targeted therapies, and for the administration of reduced doses of drug, resulting in benefits on health balance in terms of economic savings.

In 2018, our laboratory received 16,083 requests for PCT concentration determination (Table 1), requests that, compared to the previous year, increased by about 15% (total PCT 2017: 13,901). Our laboratory performs PCT assays on a Brahms KRYPTOR analyzer with the Brahms PCT sensitive Kryptor kit. In this study, we wanted to compare the assays obtained in this way with the ones performed on a LIAISON® XL so that we could use the LIAISON® XL instrumentation, in view of a highly automated organization. Moreover, given the importance of using PCT as a tool to monitor antibiotic treatment efficacy, it is important to have diagnostic systems and tests within the same hospital structure that give comparable results because they have the same calibration so that they can be used interchangeably.

Table 1. Departments that requested procalcitonin (PCT).

PCT Requests by Department		
Departments	**PCT Requests**	**%**
Long-term Hospitalization	3756	23%
Intensive Care Unit	1906	12%
Medical Immunology Department	1394	9%
Internal Medicine	1272	8%
Surgery	1003	6%
Medical Therapy Department	975	6%
Cardiology	923	6%
Hematology	908	6%
Pneumology	680	4%
Emergency Medicine–Emergency Room (ER)	675	4%
Orthopedy	630	4%
Nephrology	459	3%
Medical Oncology	408	3%
Others	397	2%
Urology	283	2%
Transplants	263	2%
Neurology	151	1%

2. Materials and Methods

From January to May 2017, we processed serum samples from 65 women (age: 22–98 years) and 106 men (age: 16–97 years), for a total of 171 subjects. The study was performed on leftover samples that were completely anonymized and de-identified, thus no informed consent was required. For the analysis, we used the LIAISON® XL analyzer (DiaSorin SpA, Saluggia VC, Italy), with LIAISON® BRAHMS PCT® II GEN reagents and calibrators, and the Brahms KRYPTOR analyzer (Brahms, Hennigsdorf, Germany distributed in Italy by Dasit), with Brahms PCT sensitive Kryptor reagents and calibrators (code M8410000, code MG611PCTK).

The LIAISON® BRAHMS PCT® II GEN test is a sandwich immunoassay based on the principle of chemiluminescence (CLIA), in which magnetic particles (solid phase) coated with a specific monoclonal antibody and another monoclonal antibody (specific for a different epitope of procalcitonin) labeled with an isoluminol derivative are used. During the first incubation, PCT binds to the conjugated antibody. Then, the solid phase is added to the reaction; the sandwich is formed only in the presence of PCT molecules, as they are bound to both antibodies. After the second incubation, the unbound material is removed with a wash cycle. Light emission generated by the chemiluminescent reaction is measured by a photomultiplier as a relative light unit (RLU). The production of light is directly proportional to the concentration of PCT present in the sample (Figure 1a) [8].

The Brahms PCT sensitive Kryptor immunofluorescence test is instead based on the TRACE (time-resolved amplified cryptate emission) technology, which measures the signal emitted by an immune complex at a delayed time.

The TRACE technology is based on the transfer of non-radioactive energy from a donor (cage structure with a europium ion in the center (cryptate)) to an acceptor, which is part of a chemically modified photo-receptive algal protein (XL665). Both the cryptate and XL665 are conjugated with monoclonal antibodies targeted to different epitopes on the PCT molecule.

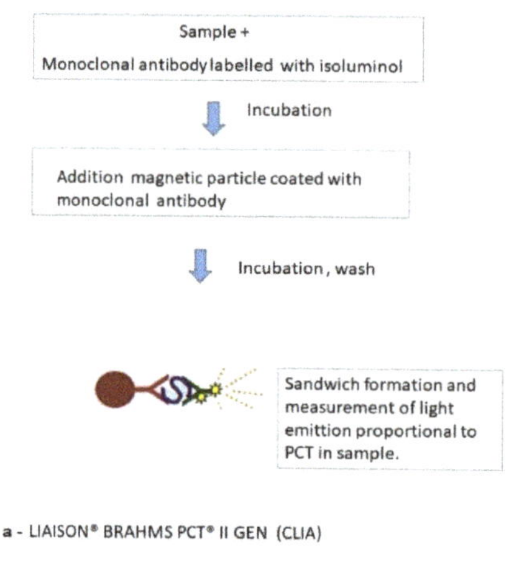

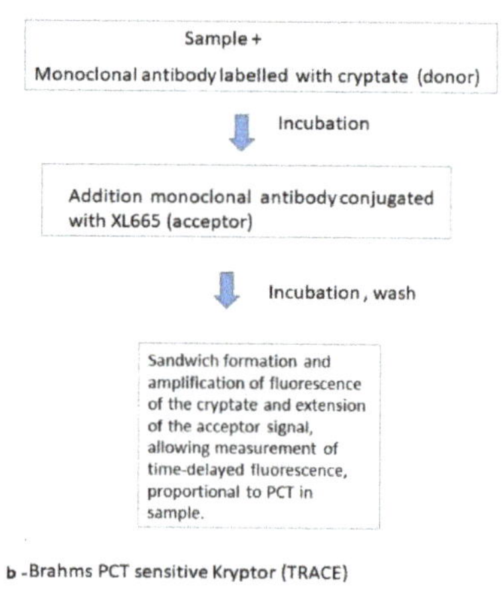

Figure 1. (**a**) Liaison Brahms PCT II Gen and (**b**) Brahms PCT sensitive Kryptor immunofluorescence test mechanism.

The proximity between the donor (cryptate) and the acceptor (XL665), when they are part of an immune complex, and the overlap between the emission spectrum of the donor and the absorption spectrum of the acceptor intensify the fluorescence signal of the cryptate, and also extend the duration of the acceptor signal. This allows measurement of the time-delayed fluorescence, which is proportional to the concentration of the analyte to be measured (Figure 1b) [9].

It should be noted that the analysis kits on both platforms allow for the use of the same monoclonal antibodies (solid phase antibodies and antibodies conjugated to XL665, respectively, in the LIAISON® BRAHMS PCT® II GEN method and the Brahms PCT sensitive Kryptor method), targeted against katacalcin, the C-terminal region of procalcitonin, which is formed by 21 amino acids). For the characteristics of reagents, see Table 2.

Table 2. Characteristics of used reagents.

	Characteristics of Used Reagents	
	LIAISON® BRAHMS PCT® GEN	**Brahms PCT Sensitive Kryptor**
Number of tests	100	100
Principle	sandwich 1 step	sandwich 1 step
Signal	Isoluminol	TRACE (time-resolved amplified cryptate emission)
Specimen	Serum, plasma	Serum, plasma
Volume of sample	100	50
Frequency of calibration	56 days	7 days
Functional Sensitivity or Limit of Quantitation (LoQ)	0.04 ng/mL	0.06 ng/mL
Limit of Detection (LoD)	0.02 ng/mL	0.02 ng/mL
Limit of Blank (LoB)		
Measuring Range ng/mL	0.02–100	0.02–50 ng/mL
Sample Dilution	Automatic	Automatic
Calibrator and Integral Stability	12 weeks	14 days
Time to first results	16 min	19 min

Regarding the statistical analysis, MedCalc 18.6 software (MedCalc bvba software, Ostend, Belgium) was used.

3. Results

With the LIAISON® BRAHMS PCT® II GEN method, 52% of the results were between 0.0 and 0.5 ng/mL, 18% between 0.5 and 2.0 ng/mL, and 30% between 2.0 and 100 ng/mL, and the maximum and minimum values were 97.2 and 0.02 ng/mL, respectively. The mean and median values were 4.09 ng/mL and 0.456 ng/mL, respectively.

With the Brahms PCT sensitive Kryptor method instead, 55% of the results were between 0.0 and 0.5 ng/mL, 21% between 0.5 and 2.0 ng/mL, and 24% between 2.0 and 100 ng/mL, and the maximum and minimum values were 103 ng/mL and 0.01 ng/mL, respectively. The mean and median values were 3.72 and 0.39 ng/mL, respectively.

There are no significant differences between the mean and the median of the results obtained with the two methods (mean 3.717 for Kryptor and 4.094 for LIAISON®; median 0.39 for Kryptor and 0.45 for LIAISON®), with the distribution of values following normality (Normal distribution: <0.001) (Figure 2A).

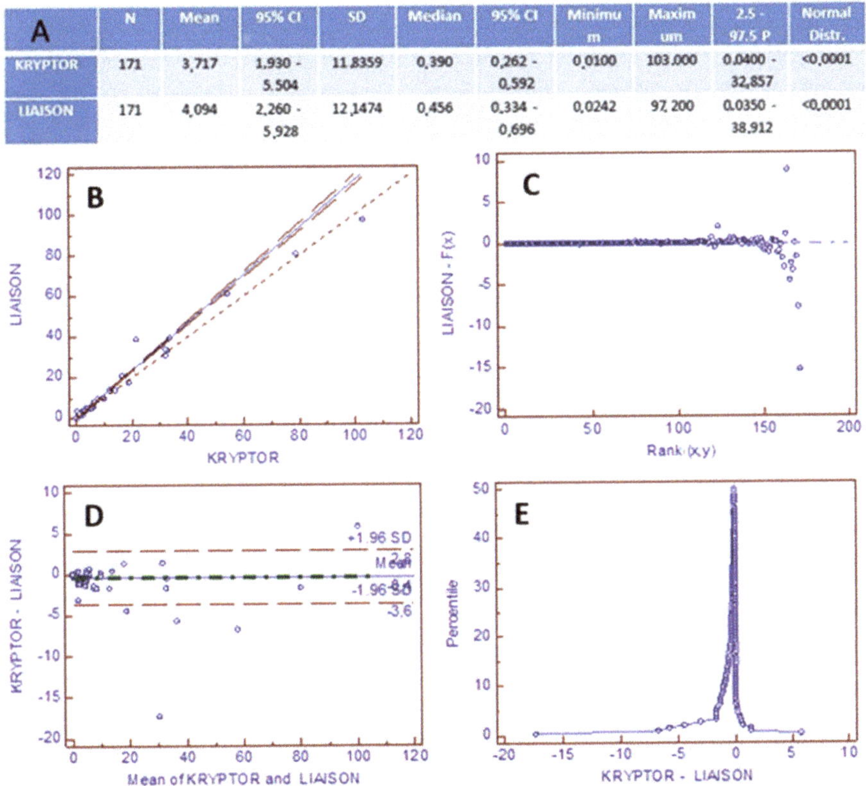

A	N	Mean	95% CI	SD	Median	95% CI	Minimum	Maximum	2.5 - 97.5 P	Normal Distr.
KRYPTOR	171	3,717	1,930 - 5,504	11,8359	0,390	0,262 - 0,592	0,0100	103,000	0,0400 - 32,857	<0,0001
LIAISON	171	4,094	2,260 - 5,928	12,1474	0,456	0,334 - 0,696	0,0242	97,200	0,0350 - 38,912	<0,0001

Figure 2. Analysis results. (**A**) Summary table of our data: mean (95% CI), median (95% CI), minimum and maximum of the two methods; (**B**) Linear regression line independent of the sample distribution Passing–Bablok (PB): $y = -0.0102086 + 1.172143x$ (Pearson coefficient = 0.99); (**C**) PB residual chart; (**D**) Comparison of Bland–Altman graph with representation of the difference, in terms of absolute value, between the two measurements shown according to the mean of the measurements; (**E**) Mountain plot graph or representation of the empirically folded cumulative distribution, obtained by calculating the distribution of percentiles relative to the differences between the two methods placed in an orderly manner.

The analysis shows an excellent correlation between the results obtained on the LIAISON®XL analyzer and on the Brahms Kryptor analyzer, with the respective reagents, both in terms of slope and intercept (Pearson coefficient: 0.99) (Figure 2B).

The regression model is further validated by the residual analysis (Figure 2C).

The Bland–Altman graph obtained by comparing the measurement differences between the two methods as a function of the average of the measurements shows a reduced dispersion of values around the mean and within the standard deviations, without significant differences (Figure 2D). The absence of significant differences is also reflected in the mountain plot graph (Figure 2E), where the graph is in fact centered on zero, missing values along the two tails.

In the literature, there is growing evidence about the fact that the determination of PCT in the monitoring of antibiotic therapy is significantly useful [6]. In this context, the PCT assay, once the appropriate antibiotic therapy has been set, allows the clinician to highlight the patient's effective response to the antibiotic, thus reducing the risk of unnecessarily prolonged treatments, the development of resistance, and other side effects.

In this regard, we have followed the course of PCT concentrations in some patients undergoing antibiotic therapy over time.

These patients, coming respectively from the emergency medicine department (patient 1, man 78 years old), a long-term care setting (patient 2, man 83 years old), and medical oncology (patient 3, woman 55 years old), were part of the population used for comparing the two methods.

The PCT determination was performed with the two assay methods throughout the antibiotic administration period, obtaining the following results.

Figures 3–5 show the trend of PCT concentrations in the three patients (PCT LIAISON® blue line; Kryptor orange line). In all three cases, the PCT values measured with LIAISON® and Kryptor are completely similar to each other, as shown by the overlap of the graphs. Furthermore, it is possible to observe a clear decrease in PCT values over time, in response to antibiotic therapy.

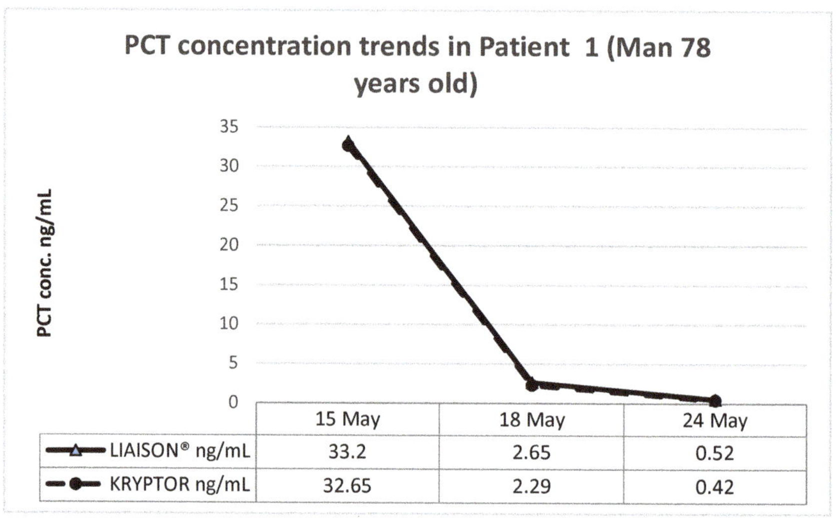

	15 May	18 May	24 May
LIAISON® ng/mL	33.2	2.65	0.52
KRYPTOR ng/mL	32.65	2.29	0.42

Figure 3. Patient 1 (Internal medicine, man 78 years old).

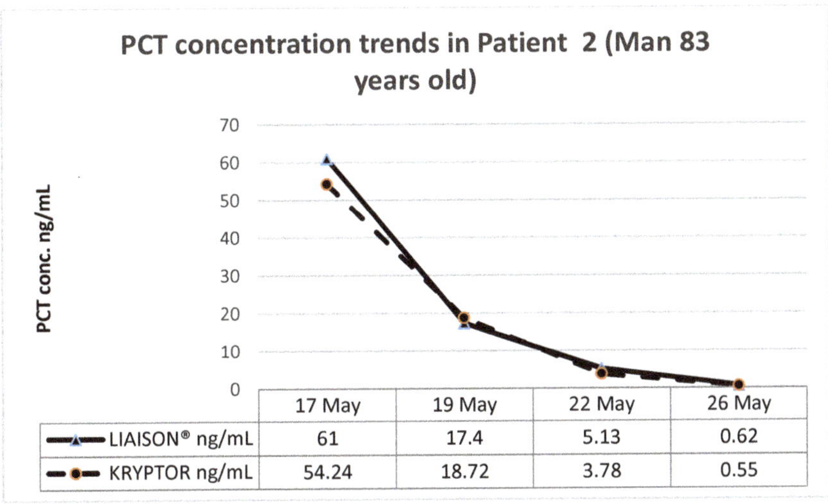

	17 May	19 May	22 May	26 May
LIAISON® ng/mL	61	17.4	5.13	0.62
KRYPTOR ng/mL	54.24	18.72	3.78	0.55

Figure 4. Patient 2 (Long-term hospitalization, man 83 years old).

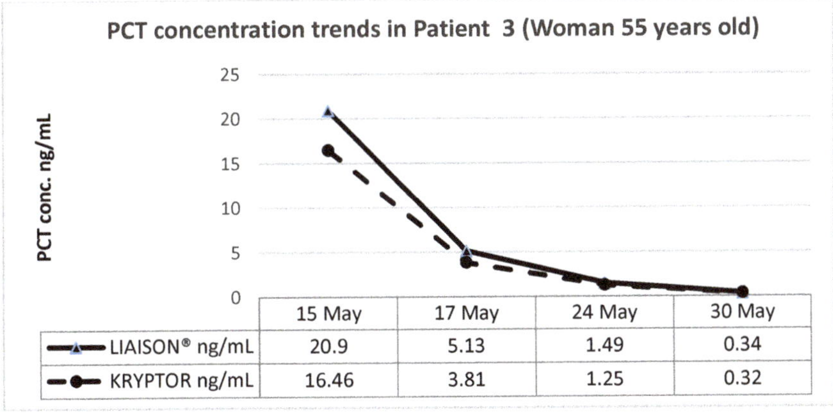

	15 May	17 May	24 May	30 May
LIAISON® ng/mL	20.9	5.13	1.49	0.34
KRYPTOR ng/mL	16.46	3.81	1.25	0.32

Figure 5. Patient 3 (Medical oncology, woman 55 years old).

4. Discussion

The determination of PCT with Brahms reagents, both on the Kryptor and LIAISON®XL platforms, presents excellent performance in terms of sensitivity and specificity. The results obtained show a high correlation between the assays performed on the two analyzers. However, the current lack of standardization of the different methods available to measure PCT remains an unmet target. Currently, PCT is determined by enzymatic, luminometric, chemiluminescent, electrochemiluminescent, fluorescent, and turbidimetric immunoassays, and the latter can be adapted for use on a large number of analytical clinical chemistry platforms, thus favoring the widespread availability of the test.

The LIAISON® BRAHMS PCT® II GEN method could be a key component for PCT-monitored antibiotic therapy and for the initial diagnosis of bacterial infection, offering good quality as well as accurate and acceptable PCT measurements. However, as with other commercially available PCT determination tests, the results must be interpreted carefully in the context of medical history, physical examination, and microbiological evaluation, given that the increase in PCT levels are not always correlated with infection and that low levels of PCT do not automatically exclude the presence of bacterial infection [10]. Therefore, the test results should not replace clinical judgment but should be integrated in a broader context, in order to obtain a better diagnostic performance [11].

Author Contributions: All the authors have contributed equally to the content of the manuscript. All the authors have accepted responsibility for the entire content of this submitted manuscript and approved submission.

Funding: None declared.

Conflicts of Interest: The authors declare no conflicts of interest.

References

1. Van der Poll, T.; van de Veerdonk, F.L.; Scicluna, B.P.; Netea, M.G. The immunopathology of sepsis and potential therapeutic targets. *Nat. Rev. Immunol.* **2017**, *17*, 407–420. [CrossRef] [PubMed]
2. Veneziani, F.; Fabrizia Petrucci Montanelli, C.; Lombardelli, L.; Anelli, M.C.; Casprini, P. Razionalizzazione e implementazione del dosaggio per la procalcitonina con metodo immunoturbidimetrico nell'USL Toscana Centro. In Proceedings of the 50° Congresso Nazionale della Società Italiana di Biochimica Clinica e Biologia Molecolare Clinica (SIBioC—Medicina di Laboratorio), Napoli, Italy, Octorber 2018; p. 252.
3. Morgenthaler, N.G.; Struck, J.; Chancerelle, Y.; Weglöhner, W.; Agay, D.; Bohuon, C.; Suarez-Domenech, V.; Bergmann, A.; Muller, B. Production of procalcitonin (PCT) in non-thyroidal tissue after LPS injection. *Horm. Metab. Res.* **2003**, *35*, 290–295. [PubMed]
4. Lippi, G. The Irreplaceable Value of Laboratory Diagnostics: Four Recent Tests that have Revolutionized Clinical Practice. *EJIFCC* **2019**, *30*, 7–13. [PubMed]
5. Tan, M.; Lu, Y.; Jiang, H.; Zhang, L. The diagnostic accuracy of procalcitonin and C-reactive protein for sepsis: A systematic review and meta-analysis. *J. Cell Biochem.* **2019**, *120*, 5852–5859. [CrossRef] [PubMed]
6. Meier, M.A.; Branche, A.; Neeser, O.L.; Wirz, Y.; Haubitz, S.; Bouadma, L.; Wolff, M.; Luyt, C.E.; Chastre, J.; Tubach, F.; et al. Procalcitonin-guided antibiotic treatment in patients with positive blood cultures: A patient-level meta-analysis of randomized trials. *Clin. Infect. Dis.* **2018**, *69*, 388–396. [CrossRef] [PubMed]
7. Zaman, S.B.; Hussain, M.A.; Nye, R.; Mehta, V.; Mamun, K.T.; Hossain, N. A Review on Antibiotic Resistance: Alarm Bells are Ringing. *Cureus* **2017**, *9*, e1403. [CrossRef] [PubMed]
8. ThermoFisher SCIENTIFIC Home Page. Available online: https://www.procalcitonin.com/pct-assays/pct-liaison.html (accessed on February 2019).
9. ThermoFisher SCIENTIFIC Home Page. Available online: https://www.procalcitonin.com/pct-assays/pct-sensitive-kryptor.html (accessed on February 2019).
10. Fortunato, A. A new sensitive automated assay for procalcitonin detection: LIAISONs BRAHMS PCTs II GEN. *Pract. Lab. Med.* **2016**, *6*, 1–7. [CrossRef] [PubMed]
11. Lippi, G.; Cervellin, G.F. Procalcitonin for diagnosing and monitoring bacterial infections: For or against? *Clin. Chem. Lab. Med.* **2018**, *56*, 1193–1195. [CrossRef] [PubMed]

Review

Capillary-Driven Flow Microfluidics Combined with Smartphone Detection: An Emerging Tool for Point-of-Care Diagnostics

Sammer-ul Hassan [1,2,*], Aamira Tariq [3], Zobia Noreen [3], Ahmed Donia [3], Syed Z. J. Zaidi [4], Habib Bokhari [3] and Xunli Zhang [1,2,*]

1 Bioengineering Research Group, Faculty of Engineering and Physical Sciences, University of Southampton, Southampton SO17 1BJ, UK
2 Institute for Life Sciences, University of Southampton, Southampton SO17 1BJ, UK
3 Department of Biosciences, Comsats University Islamabad Campus, Islamabad, Pakistan; aamira_tariq@comsats.edu.pk (A.T.); zobia_noreen@comsats.edu.pk (Z.N.); ahmeddonia123@yahoo.com (A.D.); habib@comsats.edu.pk (H.B.)
4 Institute of Chemical Engineering and Technology, University of the Punjab, Lahore, Pakistan; zohaib.icet@pu.edu.pk
* Correspondence: s.hassan@soton.ac.uk (S.H.); xl.zhang@soton.ac.uk (X.Z.)

Received: 27 June 2020; Accepted: 20 July 2020; Published: 22 July 2020

Abstract: Point-of-care (POC) or near-patient testing allows clinicians to accurately achieve real-time diagnostic results performed at or near to the patient site. The outlook of POC devices is to provide quicker analyses that can lead to well-informed clinical decisions and hence improve the health of patients at the point-of-need. Microfluidics plays an important role in the development of POC devices. However, requirements of handling expertise, pumping systems and complex fluidic controls make the technology unaffordable to the current healthcare systems in the world. In recent years, capillary-driven flow microfluidics has emerged as an attractive microfluidic-based technology to overcome these limitations by offering robust, cost-effective and simple-to-operate devices. The internal wall of the microchannels can be pre-coated with reagents, and by merely dipping the device into the patient sample, the sample can be loaded into the microchannel driven by capillary forces and can be detected via handheld or smartphone-based detectors. The capabilities of capillary-driven flow devices have not been fully exploited in developing POC diagnostics, especially for antimicrobial resistance studies in clinical settings. The purpose of this review is to open up this field of microfluidics to the ever-expanding microfluidic-based scientific community.

Keywords: microfluidics; point-of-care diagnostics; antimicrobial resistance; lab-on-a-chip; capillary-driven flow; capillary action; detections; smartphone imaging

1. Introduction

Point-of-care (POC) diagnostics enable rapid diagnosis and monitoring of the patient, e.g., at the clinic, in the field or even at home and provides on-site results to the operator instantly. POC diagnostics must provide quick and simple-to-operate testing without the requirements of laboratory technicians or sophisticated instruments [1]. Typical examples of POC devices include glucose monitoring devices and lateral flow test strips, which are widely available in the market [2]. The POC market is expanding worldwide and is predicted to be worth USD 46.7 billion in 2024 [3]. As their key features, POC devices must be portable, simple to fabricate and operate [4–6], accurate and provide results within minutes or hours, leading to the well-informed clinical decisions. These allow earlier knowledgeable treatments providing better chances to improve the health of patients as compared to the less ill-informed procedures performed while waiting for the clinical laboratory results [7]. One of the critical examples of this type of treatments is the use of broad-spectrum antibiotics for the treatment of bacterial infections where the results can take a few days to weeks to come from central clinical laboratories [8].

Microfluidics refers to the study of fluidics in micrometre channel dimensions and has been significantly developed for the miniaturization of chemical and biochemical assays offering robust, cost-effective and sensitive assays [9]. In general, these systems have been developed to provide high-throughput, parallel, and automated chemical analyses. Microfluidics has played a crucial role in the development of POC diagnostics by offering smaller sample volume consumptions and combining multiple sample-processing steps into a single device [10]. The commercialization of microfluidic devices has been attempted for various quantification platforms, including tests for numerous small molecules, protein biomarkers, nucleic acids and even cells. However, the commercialization of microfluidic-based POC devices has not been entirely successful, largely due to the requirement of handling expertise and complex miniaturized fluidic management systems [11].

Capillary-driven flow microfluidics, on the other hand, is a type of microfluidics which works on the principle of capillary action that allows the movement of fluids in capillaries or microchannels without the requirement of external pumping mechanisms [12]. This type of microfluidics offers the possibility of pre-coating of reagents into the microchannels, bringing forward ready-to-use devices for POC diagnostics ever promised by the scientific community for decades. The ideal type of capillary-driven flow-based device for POC diagnostics must provide results within a few hours or even minutes while the patient remains in the clinic, and on-spot diagnostic decisions can be made almost instantly. This review is mainly focused on the capillary-driven flow dynamics and current emerging devices for POC diagnostics, especially in tackling antimicrobial resistance studies via portable/smartphone-based devices.

2. Principles of Capillary-Driven Flow Microfluidics

Capillary-driven flow microfluidics works on the principle of capillary action that allows the movement of fluids in capillaries/microchannels without the requirement of pumps/external pressure. The liquid flows into the capillaries due to the surface tension and the wetting properties of the capillary, which overcome the effect of gravity and viscosity of the liquid. The factors which affect the liquid rise in capillary include the surface properties of the capillary, the types of fluid and the internal geometry of the capillary. Capillary-driven flow is continuous when initiated and stops at the filling of the capillary without the requirement of external control, generating a fully autonomous liquid handling system [13].

Initially, the kinetics of liquid flow along capillary were mathematically determined by the Lucas–Washburn equation, relating the capillary pressure with the fluid viscosity pressure drop at the meniscus [14]. Here, capillary pressure is the primary source of the liquid rise in the channels, which is generated by the difference in energy of liquid and air surfaces in a capillary. The most common law for a capillary rise is known as Jurin's law [15], which relates the height of the capillary with the surface energy at the interface of the two fluids (liquid and air), the contact angle at the meniscus between two phases, capillary radius, density and gravitational acceleration on the capillary (Figure 1a), as expressed in Equation (1). The Young–Laplace equation is a widely used mathematical model (Equation (2)) that relates the capillary pressure with an internal radius of the capillary, surface tension and the dynamic contact angle at the meniscus of the liquid [16].

$$h = \frac{2\gamma \cos\theta}{\rho g r} \tag{1}$$

$$h = \frac{\gamma \cos\theta}{\rho g}\left(\frac{1}{a} + \frac{1}{b}\right) \tag{2}$$

where h is the height of the liquid rise; γ is the surface tension between the liquid and the air interface; θ is the contact angle at the interface; r is the radius of the capillary; g is the gravitational acceleration in the capillary; ρ is the density of the rising liquid; a and b are the width and depth of the capillary or microchannels, respectively (Figure 1a).

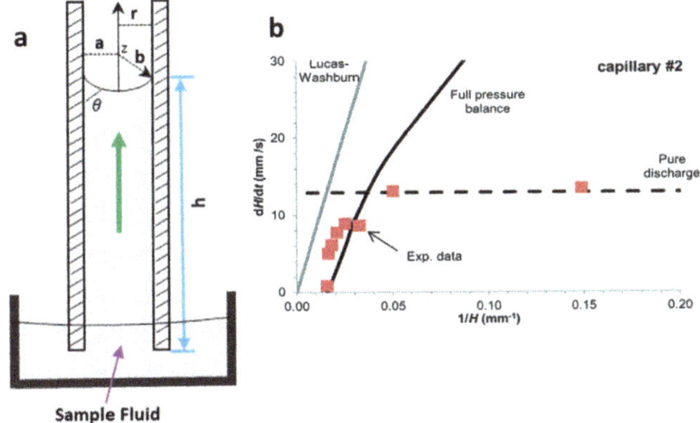

Figure 1. Capillary-driven flow dynamics. (**a**) Schematic of capillary-driven flow principle illustrating liquid rise in a capillary, height of the liquid rise (h), meniscus and contact angle at the liquid–air interface (θ) and the radius of the capillary (r); (**b**) Superficial flow velocity (dH/dt) modelling using Equation (3) predictions at full pressure balance in the capillary. The Lucas–Washburn equation is unable to predict the liquid rise in the capillary due to the absence of gravitational effect in the equation. Reproduced with permission from [16].

The Laplace pressure drop (ΔP_L) is a combination of pressure rise in the capillary and pressure drop due to the adhesiveness/friction of the surface. Reis et al. [16] modelled an equation using fluoropolymer-based capillaries to measure the superficial flow velocity (u) based on the Young–Laplace equation (Equation (3)), as shown in Figure 1b. The authors also showed that the Lucas–Washburn model was unable to propose a maximum liquid rise in the capillary because of the absence of the maximum liquid height in the modelled equation.

$$u(t) = \frac{d_h}{32\mu}\left[4\gamma cos\theta\frac{1}{h} - \rho g d_h\right] \tag{3}$$

The liquid rise in the capillaries with hydrophilic surfaces due to surface adhesiveness can be matched with the capillaries with hydrophobic surfaces, which tend to empty the capillary via gravitational forces. Figure 1b shows an example of modelling of superficial velocity based on Equation (3) for the full pressure balance. The authors predicted that the liquid rise up to 40% was driven via Laplace pressure, however, after an equilibrium Laplace height was reached, the liquid flow upward was driven by surface tension with superficial flow velocity (dH/dt) following the model Equation (3) [16].

Several materials have been used for passive pumping into microchannels, such as wool [17], cotton yarn [18], polyester [19] and paper [20]. These materials always allow fluid movement due to the hydrophilicity of the liquid sample. However, these methods provide a less stable flow rate, dead volume and the flow rates are fixed during the fabrication of these materials [21]. On the other hand, capillary-driven flow in microchannels occurs when microchannels are closed, or tubing is used, and the fluid stops upon completely filling into the microchannels. In this case, the microchannel must be significantly bigger than the width of the microchannel to initiate the flow, even if the microchannel is hydrophilic to the sample liquid. For the open microchannels, the Young–Laplace equation must be satisfied to initiate the fluid movement [22]. The flow rate in microchannels/capillaries is also not constant and is fixed during fabrication of devices but offers several benefits over materials as mentioned earlier, such as low dead volume, less contamination, and flow rate reproducibility between microchannels [22].

Capillary-driven flow has several limitations, including backflow, cross-contamination and asynchronous fluid movement inside microchannels. These problems are mainly caused by pressure differences at inlets and largely depend on the characteristics of the sample liquid [23]. Several strategies have been reported to overcome backflow problems, such as using a small flow bridge or vacuum storage [24,25]. Flow rate variability in microchannels also limits capillary flow control as mentioned above, and this problem can be circumvented by the integration of flow resistance mechanisms inside microchannels [26], varying microchannel dimensions [12], or the incorporation of porous materials at the outlets to absorb liquids [27]. Furthermore, stability and durability of capillary flow devices such as microchannels fabrication, reagent cross-linking, storage and stability must be considered for complex chemical/biochemical assays.

3. Recent Advances in Capillary-Driven Flow Microfluidics

Capillary-driven flow microfluidics has been developed predominantly for biomedical applications, e.g., numerical modelling of capillary-driven flow for polymerase chain reactions (PCR) inside capillaries was performed [28]. The authors modelled PCR in hydrophobic polydimethylsiloxane (PDMS) microchannels via computational fluid dynamics (CFD) which were also experimentally tested to compare results, as shown in Figure 2a. PCR mixture was introduced into the microchannels, and the liquid meniscus was recorded to fill all the channels in 12 s, which matched well with the modelled meniscus movements. In another development, capillary-driven flow devices were built to measure the viscosity of the fluid based on capillary action in microchannels [29,30]. A micromachined polymethyl methacrylate (PMMA) chip was used to create a capillary flow inside microchannels, and the meniscus was monitored via an automated optical system (Figure 2b) [29]. In this study, a small volume (26 μL) of the sample was loaded into the microchannels, and viscosity was measured by recording the fluid travel time between two fixed points. In contrast, Lee et al. [30] developed a capillary flow microfluidics device to measure zebrafish blood viscosity in microchannels (Figure 2c). The device was able to use small sample volumes (2 μL) and conduct continuous measurements to study the viscosity changes during embryonic development. Furthermore, the device was able to draw up fluids at higher shear rates, and the measurements were achieved in less than 2 min. Such optically enabled capillary flow systems have the capacity to be used as portable systems for POC diagnostics and analysis in chemical/pharmaceutical industries.

Polymethyl methacrylate was also used to fabricate a capillary flow device for nucleic acid biosensing applications having small microfluidic channels consisting of sealed reagent-loaded pads [31]. The system was incorporated with magnetic nanoparticles, which were released upon sample entry into the microchannel, and horseradish peroxidase (HRP) and hydrogen peroxide were mixed in a channel yielding potentiostat detectable species. This type of biosensor based on capillary-driven flow has the potential of miniaturization and commercialization for POC applications. Open microfluidics channels have also been developed for capillary flow [32], but they have disadvantages, such as sample evaporation, contamination and handling difficulties. Therefore, open microfluidics is not discussed further in this review, which is instead focused on capillary flow inside closed microchannels and capillaries.

Several biomedically relevant biomarkers were studied via capillary-driven flow microfluidics [33–38]. A capillary flow immunoassay microchip was developed by Fuchiwaki et al. [33] to quantify procollagen type I C-peptide from blood samples. Laser ablation was used to immobilize the antibody on the surface of the PMMA microchannels, which allowed the reaction via movement of liquid in the channels, as shown in the schematic (Figure 2d) [33]. The liquid sample was dropped on the channel inlets which travelled towards the antibody-coated region via capillary force, incubated and flushed out via the outlet and paper absorption. Different concentrations of the peptide (0–600 ng mL^{-1}) were calibrated by multiplexing the channels on a single platform and injecting them into parallel microchannels (Figure 2d). Enzyme inhibitor assays were also developed on capillary flow systems to enable single-step analysis and multiplex them for commercial purposes (Figure 2e) [34]. The device was made of polydimethylsiloxane (PDMS) and was named a "combinable PDMS capillary sensor". In this study, fluorescent substrates were immobilized on the membrane in the microchannels and the enzyme inhibitor solution is drawn into the microchannel via capillary force. The inhibitor solution dissolves into the membrane where the reactions occur, and detection is performed in an optically clear microchannel. This method shows the potential to multiply the assays by simply cutting long PDMS microchips coated with reagents and flowing samples in parallel. A similar PDMS capillary flow device [35] was also developed for the detection of serum biomarkers, such as glucose, potassium and alkaline phosphate assay (ALP). In this device, the PDMS was mixed with a black reagent (India ink) to render the PDMS black, and the channel surface was coated with silver layers to enhance the fluorescent signals and sensitivity of the assay (Figure 2f). The sample was introduced into the microchannels via capillary flow, and the intensities of three biomarkers were attained in parallel (Figure 2f).

Huang et al. [36] developed a capillary flow device for DNA probe detection using ordered silicon microcapillary array to control the flow of the liquid. The authors simply placed a drop of sample on the inlet (20 µL) which travelled in the microchannel with DNA probe spots (100 nL) and achieved rapid DNA detection. Recently, Soares et al. [37] developed an ultrasensitive and single-step capillary flow device for biosensing applications. The authors used bead-based technology to control the capillary flow and performed a competitive assay using only a 4.5 µL sample. The system is robust and bears a huge potential for POC diagnostics due to its turn time of 70 s per assay.

Additionally, the capillary flow has also been used for blood plasma separation in microfluidic channels [38–40], such as that developed by Madadi et al. [38], to separate plasma using 5 µL of the sample, generating a plasma of 0.1 µL for diagnostic applications. The authors used microchannel-integrated micropillars to confine red blood cells and validated the separation via the detection of thyroid-stimulating hormone. Delamarche's group [39,40] further developed plasma separation attached with immunodiagnostic devices, where C-reactive protein (CRP) was quantified by using 5 µL of human serum extracted from a blood sample and 3.6 nL of reagents solution deposited on the chip [39]. The system was able to detect low concentrations within 3 min by flowing the liquid via capillary force in the microchannels. Furthermore, a capillary flow bead-based immunoassay device for diagnostics has also been developed by the same lab [40]. These types of devices offer multiple biomedical applications with a possibility of detection via smartphones or handheld devices.

Glass capillaries, which are widely available in the market, were also utilized for capillary flow. Lapierre et al. [41] used bare glass capillaries to collect blood samples in the capillaries, as shown in Figure 2g. The authors generated profiles of blood movement in vertical capillaries, studied the effect of ageing of glass on the liquid rise and modelled the liquid rise parameters using standard equations (Equations (1)–(3)). However, fabrication of glass microchannels for POC applications is generally costly and requires tedious glass intrusions and solvent bonding. In contrast, fluoropolymer microcapillaries (FEP) have been coated with reagents to achieve hydrophilic surface inside the capillaries which can load aqueous samples rapidly (Figure 2h) [16,42–45]. Pivetal et al. [42] converted the surface of FEP microcapillaries by coating with a layer of polyvinyl alcohol (PVA) and cross-linked reagents for analyte detections (colorimetric or fluorescent), such as the detection of prostate-specific antigen (PSA) [43,44] and cytokines [45]. These FEP capillaries were injected with multiple solutions, such as PSA standard, detection antibodies, enzyme complex, washing solutions and enzymatic substrates, which were injected in all channels simultaneously. Additionally, FEP microcapillaries were placed vertically in the blood sample to draw up the liquid for ABO blood typing [16]. When the liquid rose into the capillaries, the reagents were released into the sample fluid and reacted with biomarkers to produce colour/fluorescent signal, which was detected via microscope or portable/smartphone systems. FEP microcapillaries illustrate a great potential for integration into point-of-care testing devices and offer the possibility of inexpensive biochemical analyses.

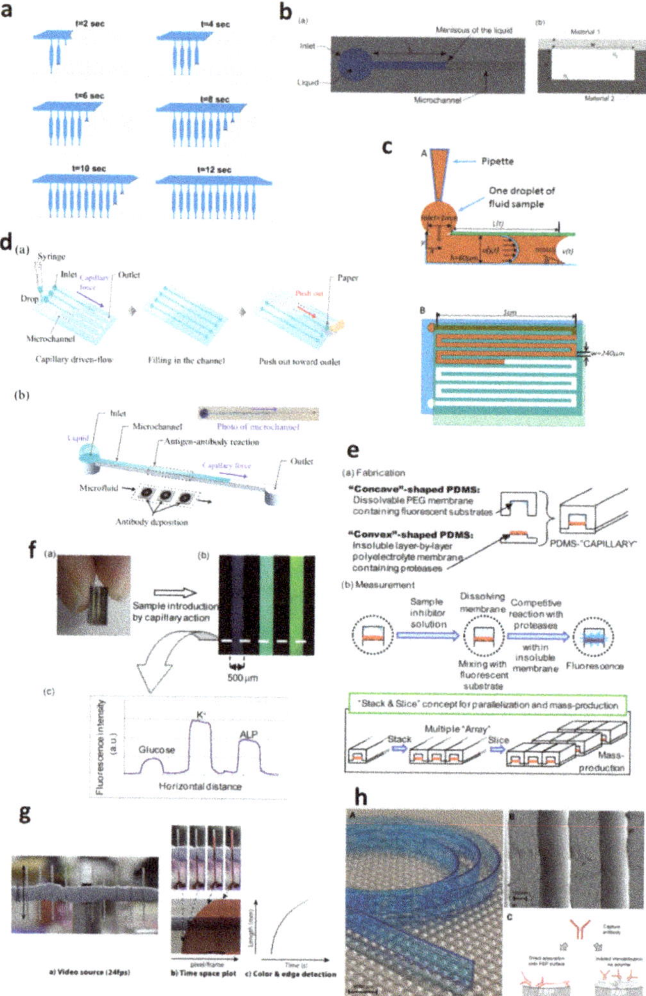

Figure 2. Applications of capillary-driven flow microfluidics. (**a**) computational fluid dynamics (CFD)-modelled filling of microchannels at multiple time points. Reproduced with permission from [28]; (**b**) Optical viscometer. Schematic of the microchannel with the capillary flow (**a**), illustration of the microchannel fabrication with two different substrates (**b**). Reproduced with permission from [29]; (**c**) Capillary flow pressure-driven zebrafish-blood-loaded microchannels. Side and top view of the sample pipetting and fluid flow in the microchannel (**A**,**B**). Reproduced with permission from [30]; (**d**) Illustration of the antibody immobilization process (**a**) with single-channel schematic (**b**). Reproduced with permission from [33]; (**e**) Combinable-polydimethylsiloxane (PDMS) capillary sensor array concept illustration. Reproduced with permission from [34]; (**f**) Combinable-PDMS capillary sensor array analysis of serum components via imaging (**a**) and fluorescence microscopy (b-c). Reproduced with permission from [35]; (**g**) Blood volume collection using glass capillaries. Capillary filling with EDTA blood (**a**) with the capillary filling plot of time−space (**b**) and color threshold levels and edge detection (**c**). Reproduced with permission from [41]; (**h**) Fluoropolymer microcapillaries illustration with flat ribbon blue dye-filled (**A**), field emission gun scanning electron microscopy (FEG-SEM) image (**B**), and capture antibody immobilization strategies (**C**). Reproduced with permission from [42].

4. Tackling Antimicrobial Resistance in Capillary-Driven Flow Devices

Microfluidics has been successfully applied for antimicrobial resistance (AMR) studies for the development of POC diagnostic devices. Still, current devices lack the simplicity and ease of operational procedures and hence are limited to laboratory settings [46–52]. Antibiotic susceptibility testing (AST) is considered to be the most widely used approach to investigate the efficacy of a single antibiotic or combination of antibiotics to find the most effective treatment for bacterial infections. Broth dilution, disk diffusion, and commercially automated systems are examples of current AST techniques [46]. These techniques have been highly standardized, yet the results are obtained after a long incubation time (15~20 h). This is due to the requirement of the bacterial population to reach the minimum detectable growth level. Scientists usually depend on measuring the optical density (OD) for sensing bacterial population with a limited detection of 10^7 colony-forming units (CFU/mL) [46,53]. Through keeping cells to a micrometer scale level, microfluidics allows individual bacterial division at early stages, which eventually lowers AST time. This could be performed within just a few hours via the indirect and direct monitoring of cell growth. Direct optical imaging is simple and can be performed on all clinical isolates. However, this is not the case in some indirect monitoring methods, due to the requisite of genetic, immunoassay modification, and complex experimental setups, meaning that some indirect methods are usually limited [46,54].

The short doubling time of bacterial cells reduces AST time. Thus, the direct observation of single-cell division within the entire bacterial population is of pivotal importance. Immobilization of bacteria on-chip is required for time-lapse growth observation of a single bacterium because of the high motility of several bacterial types. Single-cell trapping has been found to be feasible in microchambers, droplets, channels/tracks or traps [47,48,55,56]. Moreover, the confinement and capture of single *Escherichia coli* bacterium in microfluidic channels were performed in several studies with the help of dielectrophoresis (DEP) [57,58]. Under these conditions, *E. coli* AST was completed in 5 h [57] and 1 h [58] via observing the growth of single bacterium with the help of time-lapse microscopy. Agarose was also used to encapsulate single bacterium cells on-chip [49,59]. After gelation, culture media and antibiotics were added by perfusion via the gel and bacteria growth was observed for the determination of the minimal inhibitory concentrations (MICs) of several types of antibiotics on *Pseudomonas aeruginosa*, *Escherichia coli*, *Staphylococcus aureus*, *Klebsiella pneumoniae*, and *Enterococcus* spp. within only 4 h. However, scientists found tracking the growth of multilayer single bacterium cells in a 3D gel to be burdensome; therefore, the development of a polydimethylsiloxane (PDMS)–membrane–coverslip sandwich structure was required as a way to confine monolayer single bacterial cells [50,60,61]. The recent study integrated a gradient formation system to a similar chip assembly as a way for the determination of the minimum inhibitory concentration (MIC) and half-maximal inhibitory concentration (IC50) of amoxicillin on *E. coli* in just 4 h. PDMS microchannels, which were perfused with drug solutions (in the "source" channel) and culture medium (in the "sink" channel), were applied to set up a steady antibiotic concentration gradient in an agarose gel layer and antibiotic diffuses to the bacterial monolayer underneath. On the contrary to the determination of antibiotic resistance in terms of bacterial growth, monitoring cell death, which is achieved by the use of enzymatic and mechanical stress on a microfluidic chip, can facilitate rapid identification of antibiotic-resistant bacteria within 1 h [51]. Rapid AST is not the only advantage point of microfluidics. In fact, the MIC determined by microfluidic AST has been found to be comparable to those by conventional methods for standard Clinical Laboratory Standard Institute (CLSI) strains [49,52,62,63].

In contrast, capillary-driven flow microfluidics has the potential to provide robust AMR devices for POC diagnostics. Several such studies were performed; Pak et al. [64], for example, developed a latex agglutinating for the quantification of *E. coli* contamination levels in samples. In this study, agglutinates were formed via antigen binding to functionalized particles (Figure 3a). The study was performed in microchannels using reagent-coated hydrophilic adhesive tape and detecting samples in a detection zone using smartphone imaging. Reis et al. [16] developed the Lab-on-a-Stick platform for antimicrobial testing via detection in an imaging scanner, as shown in Figure 3b. The authors used FEP microcapillaries coated with reagents of various types in a single microchip and successfully used it for cell-based assays by simply dipping it into a single sample and performing ABO blood typing, microbial detection and minimum inhibitory concentrations (MIC) of antibiotics. Pivetal et al. [65] also evaluated the effect of *E. coli* concentration (urine sample) on antimicrobial susceptibility testing (AST) using a novel lab-on-a-comb platform. Multiple FEP microcapillaries were used to identify AST profiles, which were found to be in comparison with standard technologies and within the clinical range of the patient urine (103–108 CFU/mL). Recently, our group developed a Chip-and-Dip device fabricated with PMMA, and the sample loading was realized by simply dipping the chip into the sample solutions (Figure 3c) [66]. By coating microchannels with nitrocefin, it enabled beta-lactamase activity to be monitored in the microchannels. We envisage that this type of capillary flow device fabricated with cheap materials can be coated with various reagents; thus, a variety of analytes can be detected in a single chip with a multichannel microchip using suitable portable/handheld detection systems [67].

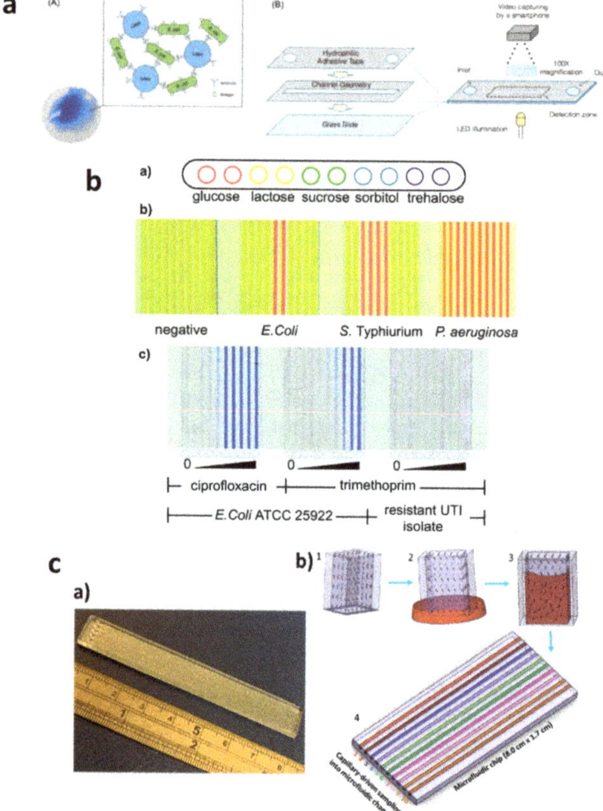

Figure 3. Applications of capillary-driven flow microfluidics for tackling antimicrobial resistance (AMR). (**a**) Immobilization of antibodies via latex agglutination test in which latex particles agglutinate upon the binding of the antibody with antigens loaded into the microchannel with the capillary flow (**A**,**B**). Reproduced with permission from [64]; (**b**) Fluoropolymer microcapillary-based identification of bacterial samples and antimicrobial susceptibility testing (**a**–**c**). Reproduced with permission from [16]; (**c**) Chip-and-dip capillary flow device fabrication and working schematic for beta-lactamase activity testing in microchannels (**a**,**b**). Reproduced with permission from [66].

5. Smartphone-Based Detection: A Crucial Component for Capillary-Driven POC Diagnostics

Optical microscopy has been routinely used for imaging and detection inside microfluidic devices and has been a critical component of lab-on-a-chip devices [68]. However, due to the bulk, expensive and expertise-dependent nature of the optical microscopy, its usage has been limited to central laboratories/hospitals [69,70]. Smartphone-based detection systems have recently emerged as alternative tools to these conventional microscopes, providing bright-field microscopy [71,72] as well as fluorescent imaging [73,74]. Smartphones are currently manufactured at much lower costs with built-in cameras, providing options for developing simple-to-operate mobile health devices [75]. Day by day, smartphones are incorporating additional features to allow data handling and signal processing and require minimal attachments to the phone's camera [76]. Several smartphone-based detection systems have been developed for biomedical applications [77–79], especially for antimicrobial resistance studies; for example, Steve et al. [80] developed a smartphone-based system for the detection of *Klebsiella pneumoniae* in 78 patients. The authors used microtiter plates for liquid handling, and a holder was 3D printed for attachment to the smartphone and placement of a microtiter plate. The system was made of an array of light-emitting diodes (LED) for sample light illumination, and an optical fiber was used for output light capturing and detection via the smartphone camera. This technique allowed faster turbidity measurements and accurate MICs and ASTs from the microwells. In another study [81], bacterial cell growth in arrays of gas-permeable microwells was studied using a smartphone-based system. The study was performed by coating microwells with antibiotics and culturing bacteria with a substrate (colorimetric) producing a signal, which was detected by a smartphone. The authors successfully obtained profiles of ASTs for uropathogens from urinary tract infection (UTI) patients. These type of studies pave the way for the miniaturization of MIC determinations and AST profiling inside microchannels, reducing sample quantities, detection intervals and handling requirements.

Capillary-driven flow devices hold a great promise in developing POC diagnostics, but still rely on accessories to detect and analyze samples and hence limit the real-time applicability of these devices in clinical settings [80–82]. Therefore, we have proposed that smartphone-based capillary flow device, which would be an ideal option for rapid image capturing and data analysis (Figure 4a,b). Several studies using smartphone-based imaging in capillary flow devices have been performed, such as the quantitation of the prostate-specific antigen in FEP microcapillaries, beta-lactamase activity in microchannels and anemia diagnosis [43,66,83]. Plevnial et al. [83] developed a standalone POC diagnostics system via integration of a 3D-printed capillary flow device with smartphone imaging, as shown in Figure 4b. The authors demonstrated the auto-mixing of reagents with blood in a capillary flow device in a minute amount of time and performed anemia diagnosis. The blood sample (5 μL) was obtained via a finger-prick method, diluted and placed in a microwell where the blood Hgb levels were detected via oxidation-reduction reaction with 3,3′,5,5′-TMB and smartphone imaging. In another study [43], a smartphone-based detection system to quantify prostate-specific antigen via colorimetric assay reaction was developed (Figure 4c). The smartphone was also converted into a fluorescent imaging microscope and hence is suitable for the study of multiple fluorescent-based assays in capillary flow systems. These types of device designs and smartphone analysis systems can provide an ideal platform for numerous disease diagnostic systems.

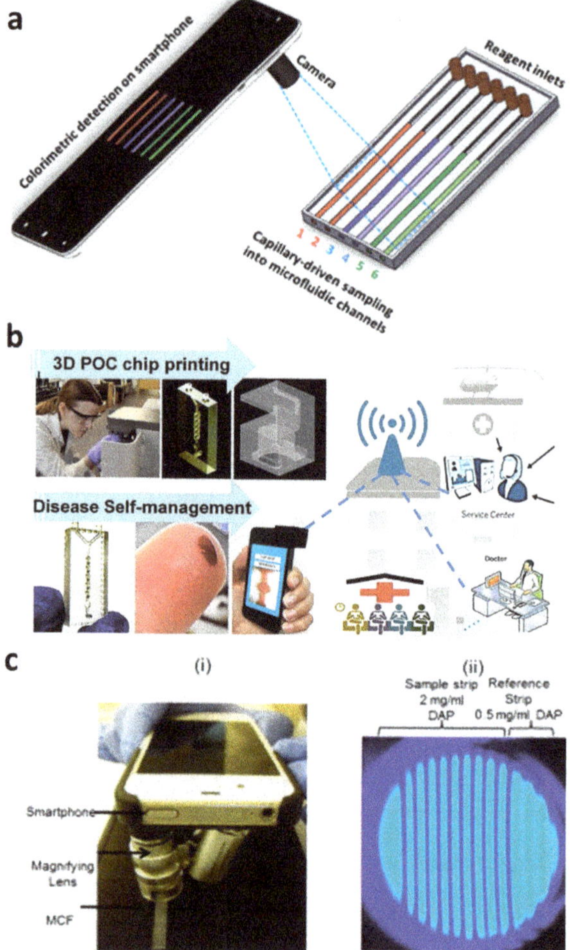

Figure 4. Smartphone-based capillary-driven flow microfluidics. (**a**) The transient filling of a microchannel at six different time points obtained using the CFD model; (**b**) smartphone-based detection system for anemia detection in microchannels via capillary flow. Reproduced with permission from [83]; (**c**) Smartphone-based fluoropolymer microcapillary detection system (**i**) and an image of fluoropolymer microcapillaries (FEP) filled with 2,3-diaminophenazine (**ii**). Reproduced with permission from [43].

6. Conclusions and Future Challenges

Point-of-care (POC) diagnostics allow the testing of patients in real-time, such as lateral flow assays widely used in the home or clinical settings. However, they have limitations, including the analysis of active bacterial/pathogenic cells, and hence are not applicable for fluid flow applications. Microfluidics plays a vital role in the development of miniaturized POC devices, but handling and accessory requirements make it less practical for deployment in clinical settings. Capillary-driven flow microfluidics has emerged as an alternative to these cumbersome techniques and offers robust, cost-effective and simple-to-operate microfluidic devices. This type of microfluidics has potential for adaptation towards point-of-care diagnostics, especially for tackling antimicrobial resistance, and requires global scientific attention. Such devices can be pre-coated with specific reagents and multiplexed for multiple analyte detections at minimum costs and fewer labour requirements. Furthermore, the elimination of pumping and fluid control systems makes this technology suitable for smartphone/portable detection, as the devices can be simply inserted into the optical detection systems holders attached to smartphones.

One of the most prominent challenges for capillary-driven flow microfluidics is the stability and control of flow rates in microchannels. The flow rates largely depend on inlet pressures and are fixed during fabrication of microchannels. Several device modifications have been attempted to overcome this challenge, such as by varying the dimensions of microchannels to control fluid movement [12] and the incorporation of porous materials at the outlets to absorb the fluid and increase micropumping time [27]. Furthermore, liquid evaporation may also occur after capillary fill and depends on the microchannel dimensions and temperature of the devices [84].

Additionally, durability and stability of microchannels and chemicals also affect the micropumping capability of the capillary flow device, such as reagent degradation over time due to light exposure, moisture or temperature. To circumvent this, the printing of chemicals, freeze-drying chemicals into microchannels, laminations or vacuum sealing can be used [85–87].

This article has provided insight for researchers working in the fields of diagnostics to gain an understanding of the capillary-driven flow microfluidics combined with smartphone/portable detections and to familiarize themselves with its potentials in developing point-of-care diagnostic devices. To gain confidence in clinical settings and break through in commercialization requires a reduction in assay steps, no use of sophisticated instruments and less handling expertise, along with highly scalable and inexpensive fabrications. We believe that capillary-driven flow microfluidics holds a true potential in bringing the lab-on-a-chip devices outside the laboratory setting as standalone point-of-care diagnostic platforms.

Funding: We thank the funding from the Economic and Social Research Council UK (ES/S000208/1). HB would like to thank the Higher Education Commission for providing funding to support this work under the grant number NRPU-3470.

Acknowledgments: The presented study was completed as part of the DOSA Project (Diagnostics for One Health and User Driven Solutions for AMR, www.dosa-diagnostics.org). DOSA is funded by UK Research and Innovation/Economic Social Science Research Council, Newton Fund, and the Government of India's Department of Biotechnology. We thank the funding from the Economic and Social Research Council UK (ES/S000208/1).

Conflicts of Interest: The authors declare no conflict of interest.

References

1. Nguyen, T.; Andreasen, S.Z.; Wolff, A.; Bang, D.D. From lab on a chip to point of care devices: The role of open source microcontrollers. *Micromachines* **2018**, *9*, 403. [CrossRef] [PubMed]
2. Koczula, K.M.; Gallotta, A. Lateral flow assays. *Essays Biochem.* **2016**. [CrossRef]
3. Point-of-Care Diagnostics Market. Available online: https://www.marketsandmarkets.com/Market-Reports/point-of-care-diagnostic-market-106829185.html (accessed on 2 June 2020).
4. Nguyen, T.; Vinayaka, A.C.; Bang, D.D.; Wolff, A. A complete protocol for rapid and low-cost fabrication of polymer microfluidic chips containing three-dimensional microstructures used in point-of-care devices. *Micromachines* **2019**, *10*, 624. [CrossRef]
5. Nguyen, T.; Chidambara, V.A.; Bang, D.D.; Wolff, A. Large-scale fabrication of microfluidic chips with three-dimensional microstructures for point of care application. In Proceedings of the 23rd International Conference on Miniaturized Systems for Chemistry and Life Sciences, Basel, Switzerland, 27–31 October 2019.
6. Hassan, S.U.; Nightingale, A.M.; Niu, X. Micromachined optical flow cell for sensitive measurement of droplets in tubing. *Biomed. Microdevices* **2018**. [CrossRef]
7. Liu, Z.; Banaei, N.; Ren, K. Microfluidics for Combating Antimicrobial Resistance. *Trends Biotechnol.* **2017**. [CrossRef]
8. Hassan, S.; Zhang, X. Microfluidics as an Emerging Platform for Tackling Antimicrobial Resistance (AMR): A Review. *Curr. Anal. Chem.* **2019**, *15*. [CrossRef]
9. Whitesides, G.M. The origins and the future of microfluidics. *Nature* **2006**, *442*, 368–373. [CrossRef]

10. St John, A.; Price, C.P. Existing and Emerging Technologies for Point-of-Care Testing. *Clin. Biochem. Rev.* **2014**, *35*, 155–167.
11. Chin, C.D.; Linder, V.; Sia, S.K. Commercialization of microfluidic point-of-care diagnostic devices. *Lab Chip* **2012**, *12*, 2118–2134. [CrossRef] [PubMed]
12. Olanrewaju, A.; Beaugrand, M.; Yafia, M.; Juncker, D. Capillary microfluidics in microchannels: From microfluidic networks to capillaric circuits. *Lab Chip* **2018**. [CrossRef]
13. Stange, M.; Dreyer, M.E.; Rath, H.J. Capillary driven flow in circular cylindrical tubes. *Phys. Fluids* **2003**. [CrossRef]
14. Masoodi, R.; Pillai, K.; Masoodi, R.; Pillai, K. Traditional Theories of Wicking. In *Wicking in Porous Materials*; Taylor & Francis: Abingdon, UK, 2012.
15. Batchelor, G.K. *An Introduction to Fluid Dynamics*; Cambridge University Press: Cambridge, UK, 2000.
16. Reis, N.M.; Pivetal, J.; Loo-Zazueta, A.L.; Barros, J.M.S.; Edwards, A.D. Lab on a stick: Multi-analyte cellular assays in a microfluidic dipstick. *Lab Chip* **2016**. [CrossRef] [PubMed]
17. Jeon, S.H.; Hwang, K.H.; Jung, W.S.; Seo, H.J.; Nam, S.W.; Boo, J.H.; Yun, S.H. Flow manipulation in thread-based microfluidics by tuning the wettability of wool. *J. Biomed. Nanotechnol.* **2015**, *11*, 319–324. [CrossRef]
18. Safavieh, R.; Zhou, G.Z.; Juncker, D. Microfluidics made of yarns and knots: From fundamental properties to simple networks and operations. *Lab Chip* **2011**. [CrossRef]
19. Lynn, N.S.; Dandy, D.S. Passive microfluidic pumping using coupled capillary/evaporation effects. *Lab Chip* **2009**. [CrossRef]
20. Yamada, K.; Shibata, H.; Suzuki, K.; Citterio, D. Toward practical application of paper-based microfluidics for medical diagnostics: State-of-the-art and challenges. *Lab Chip* **2017**. [CrossRef] [PubMed]
21. Berejnov, V.; Djilali, N.; Sinton, D. Lab-on-chip methodologies for the study of transport in porous media: Energy applications. *Lab Chip* **2008**. [CrossRef]
22. Casavant, B.P.; Berthier, E.; Theberge, A.B.; Berthier, J.; Montanez-Sauri, S.I.; Bischel, L.L.; Brakke, K.; Hedman, C.J.; Bushman, W.; Keller, N.P.; et al. Suspended microfluidics. *Proc. Natl. Acad. Sci. USA* **2013**. [CrossRef]
23. Lee, Y.; Seder, I.; Kim, S.J. Influence of surface tension-driven network parameters on backflow strength. *RSC Adv.* **2019**. [CrossRef]
24. Kim, S.J.; Lim, Y.T.; Yang, H.; Shin, Y.B.; Kim, K.; Lee, D.S.; Park, S.H.; Kim, Y.T. Passive microfluidic control of two merging streams by capillarity and relative flow resistance. *Anal. Chem.* **2005**. [CrossRef]
25. Zhai, Y.; Wang, A.; Koh, D.; Schneider, P.; Oh, K.W. A robust, portable and backflow-free micromixing device based on both capillary- and vacuum-driven flows. *Lab Chip* **2018**. [CrossRef] [PubMed]
26. Zhang, X.; Wang, X.; Chen, K.; Cheng, J.; Xiang, N.; Ni, Z. Passive flow regulator for precise high-throughput flow rate control in microfluidic environments. *RSC Adv.* **2016**. [CrossRef]
27. Rich, M.; Mohd, O.; Ligler, F.S.; Walker, G.M. Characterization of glass frit capillary pumps for microfluidic devices. *Microfluid. Nanofluidics* **2019**. [CrossRef]
28. Ramalingam, N.; Warkiani, M.E.; Ramalingam, N.; Keshavarzi, G.; Hao-Bing, L.; Hai-Qing, T.G. Numerical and experimental study of capillary-driven flow of PCR solution in hybrid hydrophobic microfluidic networks. *Biomed. Microdevices* **2016**. [CrossRef]
29. Bamshad, A.; Nikfarjam, A.; Sabour, M.H. Capillary-based micro-optofluidic viscometer. *Meas. Sci. Technol.* **2018**. [CrossRef]
30. Lee, J.; Chou, T.C.; Kang, D.; Kang, H.; Chen, J.; Baek, K.I.; Wang, W.; DIng, Y.; Di Carlo, D.I.; Tai, Y.C.; et al. A Rapid Capillary-Pressure Driven Micro-Channel to Demonstrate Newtonian Fluid Behavior of Zebrafish Blood at High Shear Rates. *Sci. Rep.* **2017**. [CrossRef]
31. He, F.; Wang, Y.; Jin, S.; Nugen, S.R. Development of a capillary-driven, microfluidic, nucleic acid biosensor. *Sens. Transducers* **2011**, *13*, 150–158.
32. Lade, R.K.; Hippchen, E.J.; Macosko, C.W.; Francis, L.F. Dynamics of Capillary-Driven Flow in 3D Printed Open Microchannels. *Langmuir* **2017**. [CrossRef]
33. Fuchiwaki, Y.; Takaoka, H. UV-laser-assisted modification of poly(methyl methacrylate) and its application to capillary-driven-flow immunoassay. *J. Micromech. Microeng.* **2015**. [CrossRef]

34. Uchiyama, Y.; Okubo, F.; Akai, K.; Fujii, Y.; Henares, T.G.; Kawamura, K.; Yao, T.; Endo, T.; Hisamoto, H. Combinable poly(dimethyl siloxane) capillary sensor array for single-step and multiple enzyme inhibitor assays. *Lab Chip* **2012**. [CrossRef]
35. Fujii, Y.; Henares, T.G.; Kawamura, K.; Endo, T.; Hisamoto, H. Bulk- and surface-modified combinable PDMS capillary sensor array as an easy-to-use sensing device with enhanced sensitivity to elevated concentrations of multiple serum sample components. *Lab Chip* **2012**. [CrossRef] [PubMed]
36. Huang, C.; Jones, B.J.; Bivragh, M.; Jans, K.; Lagae, L.; Peumans, P. A capillary-driven microfluidic device for rapid DNA detection with extremely low sample consumption. In Proceedings of the 17th International Conference on Miniaturized Systems for Chemistry and Life Sciences MicroTAS 2013, Freiburg, Germany, 27–31 October 2013.
37. Epifania, R.; Soares, R.R.G.; Pinto, I.F.; Chu, V.; Conde, J.P. Capillary-driven microfluidic device with integrated nanoporous microbeads for ultrarapid biosensing assays. *Sens. Actuators B Chem.* **2018**. [CrossRef]
38. Madadi, H.; Casals-Terré, J.; Mohammadi, M. Self-driven filter-based blood plasma separator microfluidic chip for point-of-care testing. *Biofabrication* **2015**. [CrossRef] [PubMed]
39. Gervais, L.; Delamarche, E. Toward one-step point-of-care immunodiagnostics using capillary-driven microfluidics and PDMS substrates. *Lab Chip* **2009**. [CrossRef]
40. Delamarche, E.; Temiz, Y.; Gökçe, O.; Arango, Y. Precision Diagnostics for Mobile Health Using Capillary-driven Microfluidics. *Chim. Int. J. Chem.* **2017**. [CrossRef] [PubMed]
41. Lapierre, F.; Gooley, A.; Breadmore, M. Principles around Accurate Blood Volume Collection Using Capillary Action. *Langmuir* **2017**. [CrossRef]
42. Pivetal, J.; Pereira, F.M.; Barbosa, A.I.; Castanheira, A.P.; Reis, N.M.; Edwards, A.D. Covalent immobilisation of antibodies in Teflon-FEP microfluidic devices for the sensitive quantification of clinically relevant protein biomarkers. *Analyst* **2017**. [CrossRef]
43. Barbosa, A.I.; Gehlot, P.; Sidapra, K.; Edwards, A.D.; Reis, N.M. Portable smartphone quantitation of prostate specific antigen (PSA) in a fluoropolymer microfluidic device. *Biosens. Bioelectron.* **2015**. [CrossRef]
44. Barbosa, A.I.; Castanheira, A.P.; Edwards, A.D.; Reis, N.M. A lab-in-a-briefcase for rapid prostate specific antigen (PSA) screening from whole blood. *Lab Chip* **2014**. [CrossRef]
45. Castanheira, A.P.; Barbosa, A.I.; Edwards, A.D.; Reis, N.M. Multiplexed femtomolar quantitation of human cytokines in a fluoropolymer microcapillary film. *Analyst* **2015**. [CrossRef]
46. Pulido, M.R.; García-Quintanilla, M.; Martín-Peña, R.; Cisneros, J.M.; McConnell, M.J. Progress on the development of rapid methods for antimicrobial susceptibility testing. *J. Antimicrob. Chemother.* **2013**. [CrossRef] [PubMed]
47. Dai, J.; Yoon, S.H.; Sim, H.Y.; Yang, Y.S.; Oh, T.K.; Kim, J.F.; Hong, J.W. Charting microbial phenotypes in multiplex nanoliter batch bioreactors. *Anal. Chem.* **2013**. [CrossRef] [PubMed]
48. Long, Z.; Nugent, E.; Javer, A.; Cicuta, P.; Sclavi, B.; Cosentino Lagomarsino, M.; Dorfman, K.D. Microfluidic chemostat for measuring single cell dynamics in bacteria. *Lab Chip* **2013**. [CrossRef] [PubMed]
49. Choi, J.; Yoo, J.; Lee, M.; Kim, E.G.; Lee, J.S.; Lee, S.; Joo, S.; Song, S.H.; Kim, E.C.; Lee, J.C.; et al. A rapid antimicrobial susceptibility test based on single-cell morphological analysis. *Sci. Transl. Med.* **2014**. [CrossRef]
50. Golchin, S.A.; Stratford, J.; Curry, R.J.; McFadden, J. A microfluidic system for long-term time-lapse microscopy studies of mycobacteria. *Tuberculosis* **2012**. [CrossRef]
51. Kalashnikov, M.; Lee, J.C.; Campbell, J.; Sharon, A.; Sauer-Budge, A.F. A microfluidic platform for rapid, stress-induced antibiotic susceptibility testing of Staphylococcus aureus. *Lab Chip* **2012**. [CrossRef]
52. Li, B.; Qiu, Y.; Glidle, A.; McIlvenna, D.; Luo, Q.; Cooper, J.; Shi, H.C.; Yin, H. Gradient microfluidics enables rapid bacterial growth inhibition testing. *Anal. Chem.* **2014**. [CrossRef]
53. Dalgaard, P.; Ross, T.; Kamperman, L.; Neumeyer, K.; McMeekin, T.A. Estimation of bacterial growth rates from turbidimetric and viable count data. *Int. J. Food Microbiol.* **1994**. [CrossRef]
54. Dai, J.; Hamon, M.; Jambovane, S. Microfluidics for antibiotic susceptibility and toxicity testing. *Bioengineering* **2016**, 3, 25. [CrossRef]
55. Shemesh, J.; Arye, T.B.; Avesar, J.; Kang, J.H.; Fine, A.; Super, M.; Meller, A.; Ingber, D.E.; Levenberg, S. Stationary nanoliter droplet array with a substrate of choice for single adherent/nonadherent cell incubation and analysis. *Proc. Natl. Acad. Sci. USA* **2014**. [CrossRef]

56. Rowat, A.C.; Bird, J.C.; Agresti, J.J.; Rando, O.J.; Weitz, D.A. Tracking lineages of single cells in lines using a microfluidic device. *Proc. Natl. Acad. Sci. USA* **2009**. [CrossRef] [PubMed]
57. Peitz, I.; van Leeuwen, R. Single-cell bacteria growth monitoring by automated DEP-facilitated image analysis. *Lab Chip* **2010**. [CrossRef] [PubMed]
58. Lu, Y.; Gao, J.; Zhang, D.D.; Gau, V.; Liao, J.C.; Wong, P.K. Single cell antimicrobial susceptibility testing by confined microchannels and electrokinetic loading. *Anal. Chem.* **2013**. [CrossRef]
59. Choi, J.; Jung, Y.G.; Kim, J.; Kim, S.; Jung, Y.; Na, H.; Kwon, S. Rapid antibiotic susceptibility testing by tracking single cell growth in a microfluidic agarose channel system. *Lab Chip* **2013**. [CrossRef] [PubMed]
60. Wong, I.; Atsumi, S.; Huang, W.C.; Wu, T.Y.; Hanai, T.; Lam, M.L.; Tang, P.; Yang, J.; Liao, J.C.; Ho, C.M. An agar gel membrane-PDMS hybrid microfluidic device for long term single cell dynamic study. *Lab Chip* **2010**. [CrossRef]
61. Wakamoto, Y.; Dhar, N.; Chait, R.; Schneider, K.; Signorino-Gelo, F.; Leibler, S.; McKinney, J.D. Dynamic persistence of antibiotic-stressed mycobacteria. *Science* **2013**. [CrossRef] [PubMed]
62. Hou, Z.; An, Y.; Hjort, K.; Hjort, K.; Sandegren, L.; Wu, Z. Time lapse investigation of antibiotic susceptibility using a microfluidic linear gradient 3D culture device. *Lab Chip* **2014**. [CrossRef]
63. Matsumoto, Y.; Sakakihara, S.; Grushnikov, A.; Kikuchi, K.; Noji, H.; Yamaguchi, A.; Iino, R.; Yagi, Y.; Nishino, K. A microfluidic channel method for rapid drug-susceptibility testing of Pseudomonas aeruginosa. *PLoS ONE* **2016**. [CrossRef]
64. Pak, O.S.; Pietrzyk, K.; Ly, A.; Maldonado-Liu, A.; Fukuoka, S.; Kim, U. Quantification of a latex agglutination assay for bacterial pathogen detection in a low-cost capillary-driven fluidic platform. In Proceedings of the 2016 IEEE Global Humanitarian Technology Conference (GHTC), Seattle, WA, USA, 13–16 October 2016.
65. Pivetal, J.; Woodward, M.J.; Reis, N.M.; Edwards, A.D. High Dynamic Range Bacterial Cell Detection in a Novel 'Lab-In-A-Comb' for High-Throughput Antibiotic Susceptibility Testing of Clinical Urine Samples. *bioRxiv* **2018**. [CrossRef]
66. Hassan, S.; Zhang, X. Design and Fabrication of Capillary-Driven Flow Device for Point-Of-Care Diagnostics. *Biosensors* **2020**, *10*, 39. [CrossRef]
67. Hassan, S.; Zhang, X. Design and Fabrication of Optical Flow Cell for Multiplex Detection of β-lactamase in Microchannels. *Micromachines* **2020**, *11*, 385. [CrossRef] [PubMed]
68. Huang, X.; Jiang, Y.; Liu, X.; Xu, H.; Han, Z.; Rong, H.; Yang, H.; Yan, M.; Yu, H. Machine learning based single-frame super-resolution processing for lensless blood cell counting. *Sensors* **2016**, *16*, 1836. [CrossRef] [PubMed]
69. Huang, X.; Guo, J.; Wang, X.; Yan, M.; Kang, Y.; Yu, H. A contact-imaging based microfluidic cytometer with machine-learning for single-frame super-resolution processing. *PLoS ONE* **2014**. [CrossRef] [PubMed]
70. Vashist, S.K.; Luppa, P.B.; Yeo, L.Y.; Ozcan, A.; Luong, J.H.T. Emerging Technologies for Next-Generation Point-of-Care Testing. *Trends Biotechnol.* **2015**. [CrossRef] [PubMed]
71. Switz, N.A.; D'Ambrosio, M.V.; Fletcher, D.A. Low-cost mobile phone microscopy with a reversed mobile phone camera lens. *PLoS ONE* **2014**. [CrossRef]
72. Lee, S.A.; Yang, C. A smartphone-based chip-scale microscope using ambient illumination. *Lab Chip* **2014**. [CrossRef]
73. Lee, W.I.; Shrivastava, S.; Duy, L.T.; Kim, B.Y.; Son, Y.M.; Lee, N.E. A smartphone imaging-based label-free and dual-wavelength fluorescent biosensor with high sensitivity and accuracy. *Biosens. Bioelectron.* **2017**. [CrossRef]
74. Wei, Q.; Acuna, G.; Kim, S.; Vietz, C.; Tseng, D.; Chae, J.; Shir, D.; Luo, W.; Tinnefeld, P.; Ozcan, A. Plasmonics enhanced smartphone fluorescence microscopy. *Sci. Rep.* **2017**. [CrossRef] [PubMed]
75. Huang, X.; Xu, D.; Chen, J.; Liu, J.; Li, Y.; Song, J.; Ma, X.; Guo, J. Smartphone-based analytical biosensors. *Analyst* **2018**. [CrossRef]
76. Roda, A.; Michelini, E.; Zangheri, M.; di Fusco, M.; Calabria, D.; Simoni, P. Smartphone-based biosensors: A critical review and perspectives. *TrAC Trends Anal. Chem.* **2016**. [CrossRef]
77. Xu, D.; Huang, X.; Guo, J.; Ma, X. Automatic smartphone-based microfluidic biosensor system at the point of care. *Biosens. Bioelectron.* **2018**. [CrossRef] [PubMed]
78. Liang, P.S.; Park, T.S.; Yoon, J.Y. Rapid and reagentless detection of microbial contamination within meat utilizing a smartphone-based biosensor. *Sci. Rep.* **2014**. [CrossRef] [PubMed]

79. Karlsen, H.; Dong, T. Smartphone-Based Rapid Screening of Urinary Biomarkers. *IEEE Trans. Biomed. Circuits Syst.* **2017**. [CrossRef] [PubMed]
80. Feng, S.; Tseng, D.; di Carlo, D.; Garner, O.B.; Ozcan, A. High-throughput and automated diagnosis of antimicrobial resistance using a cost-effective cellphone-based micro-plate reader. *Sci. Rep.* **2016**. [CrossRef]
81. Kadlec, M.W.; You, D.; Liao, J.C.; Wong, P.K. A Cell Phone-Based Microphotometric System for Rapid Antimicrobial Susceptibility Testing. *J. Lab. Autom.* **2014**. [CrossRef]
82. Tseng, D.; Mudanyali, O.; Oztoprak, C.; Isikman, S.O.; Sencan, I.; Yaglidere, O.; Ozcan, A. Lensfree microscopy on a cellphone. *Lab Chip* **2010**. [CrossRef]
83. Plevniak, K.; Campbell, M.; Myers, T.; Hodges, A.; He, M. 3D printed auto-mixing chip enables rapid smartphone diagnosis of anemia. *Biomicrofluidics* **2016**. [CrossRef]
84. Xu, L.; Wang, A.; Li, X.; Oh, K.W. Passive micropumping in microfluidics for point-of-care testing. *Biomicrofluidics* **2020**. [CrossRef]
85. Gubala, V.; Harris, L.F.; Ricco, A.J.; Tan, M.X.; Williams, D.E. Point of care diagnostics: Status and future. *Anal. Chem.* **2012**. [CrossRef]
86. Gootenberg, J.S.; Abudayyeh, O.O.; Kellner, M.J.; Joung, J.; Collins, J.J.; Zhang, F. Multiplexed and portable nucleic acid detection platform with Cas13, Cas12a and Csm6. *Science* **2018**. [CrossRef]
87. Yeh, E.C.; Fu, C.C.; Hu, L.; Thakur, R.; Feng, J.; Lee, L.P. Self-powered integrated microfluidic point-of-care low-cost enabling (SIMPLE) chip. *Sci. Adv.* **2017**. [CrossRef] [PubMed]

Review

Viscoelastic Hemostatic Assays: Moving from the Laboratory to the Site of Care—A Review of Established and Emerging Technologies

Jan Hartmann [1,*], Matthew Murphy [1] and Joao D. Dias [2]

1 Haemonetics Corporation, Boston, MA 02110, USA; MMurphy@Haemonetics.com
2 Haemonetics SA, Signy CH, 1274 Signy-Centre, Switzerland; joao.dias@haemonetics.com
* Correspondence: jan.hartmann@haemonetics.com; Tel.: +1-781-348-7396

Received: 28 January 2020; Accepted: 19 February 2020; Published: 21 February 2020

Abstract: Viscoelastic-based techniques to evaluate whole blood hemostasis have advanced substantially since they were first developed over 70 years ago but are still based upon the techniques first described by Dr. Hellmut Hartert in 1948. Today, the use of thromboelastography, the method of testing viscoelastic properties of blood coagulation, has moved out of the research laboratory and is now more widespread, used commonly during surgery, in emergency departments, intensive care units, and in labor wards. Thromboelastography is currently a rapidly growing field of technological advancement and is attracting significant investment. This review will first describe the history of the viscoelastic testing and the established first-generation devices, which were developed for use within the laboratory. This review will then describe the next-generation hemostasis monitoring devices, which were developed for use at the site of care for an expanding range of clinical applications. This review will then move on to experimental technologies, which promise to make viscoelastic testing more readily available in a wider range of clinical environments in the endeavor to improve patient care.

Keywords: blood; coagulation; hemostasis; point of care; ROTEM; TEG; thromboelastography; VHA; viscoelastic testing

1. Introduction

The use of techniques to evaluate hemostasis utilizing the viscoelastic properties of whole blood have been reported since the 1940s. The technology and clinical applications of viscoelastic hemostatic assays grew slowly over the next 40 years. In parallel, the use of today's standard hemostatic assays (prothrombin time, activated partial thromboplastin, international normalized ratio, etc.) became common. However, these assays only shed light on a small part of the overall hemostasis process. In the 1980s, the use of thromboelastography became more widespread during high blood loss procedures, such as cardiac surgery and liver transplantation [1,2], and in the management of major bleeding [3]. More recently, instrumentation quality and ease of use has improved dramatically. Viscoelastic hemostatic monitoring is now a rapidly growing field, drawing significant attention and investment. The current technological landscape has a host of new technologies being developed to improve the capability of the instrumentation and bring it closer to the site of care.

2. What Is Viscoelastic Testing?

Viscoelasticity is the characteristic of a material that behaves in both a viscous (permanent deformation) and elastic (temporary deformation) manner. Prior to clotting, whole blood is a purely viscous material and the response to shear stress will be permanent deformation. Once the shear stress is removed, the sample will not return to its original shape. As a blood sample begins to clot, the fluid becomes less viscous and more elastic in nature. The deformation induced by the shear stress becomes temporary as the fluid tries to return to its original shape. The primary measurement during viscoelastic hemostatic monitoring is the observation of the transition from a viscous to elastic state, and the measurement of the shear elastic modulus: a measure of the amount of force required to shear a material (also known as shear stiffness). The elastic modulus is defined as the ratio of shear stress to shear strain (Figure 1).

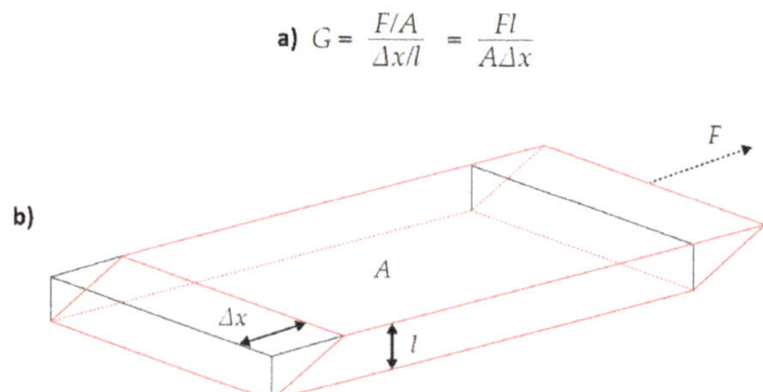

Figure 1. Shear elastic modulus. (**a**) Mathematical formula to express shear modulus; (**b**) Schematic representation of the shear principle. G = shear modulus; F = force; A = area; F/A = shear stress; Δx = transverse displacement; l = initial length. Reproduced with permission from [4].

3. History of Viscoelastic Testing

Viscoelastic hemostatic monitoring was first described in 1948 by Dr. Hellmut Hartert in Germany [5,6]. Dr. Hartert wanted a mechanism to quantify the dynamics of blood clot formation. He developed a mechanism that consisted of a cup with a concentric pin suspended within (Figure 2). The pin was suspended with a thin steel wire with a diameter of 0.2 mm, which acted as a torsional spring.

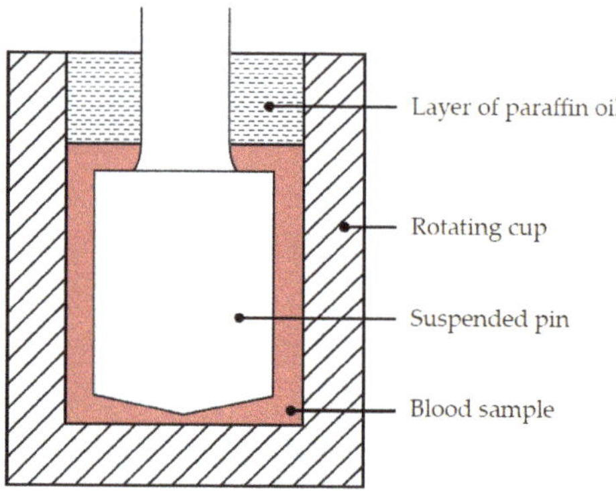

Figure 2. Dr. Hartert's cup and pin mechanism. Schematic drawing representing parts of the thromboelastograph that were in direct contact with the blood sample. The rotating cup was approximately 8 x 12 mm and made from stainless steel, the surface of which prevented detachment of the blood clot during cup rotation. The blood sample was covered by a layer of paraffin oil to prevent evaporation of the sample. Reproduced with permission from [4].

To perform a test, an activated sample of blood is placed in the cup and the pin lowered into place. The cup rotates in each direction by 1/24 radian, or 1/12 radian total rotation. The rotation occurs slowly, taking 3.5 s for one direction. The cup then comes to a stop for 1 s and then moves back in the other direction at the same speed. An entire "cycle" takes 9 s (two motion periods and two stationary periods). As the cup rotates with a viscous material (whole blood), the pin does not move. The shear between the rotating cup and stationary pin results in a permanent shear deformation of the blood. As the clot forms and grows in strength, the fluid within the cup begins its transition from a viscous to an elastic state. Energy is stored within the elasticity of the clot and the clot will try to return to its original shape, exerting a force on the pin, which causes the pin to rotate on its axis. The small rotations of the pin are transmitted to a film via a mirror coupled to the pin, which is illuminated by a slit lamp (Figure 3). The movement of the cup and pin after clotting is represented as a graphical chart in Figure 4.

Figure 3. Output from Dr Hartert's cup and pin. Representation of the output from the entire cycle of the cup and pin system. The R period was described as the reaction time, g as the growth of the clot and s as the stable period clot strength. The amplitude of the waveform is proportional to the shear modulus of the clot within its elastic region and is analogous to clot strength. Reproduced with permission from [5].

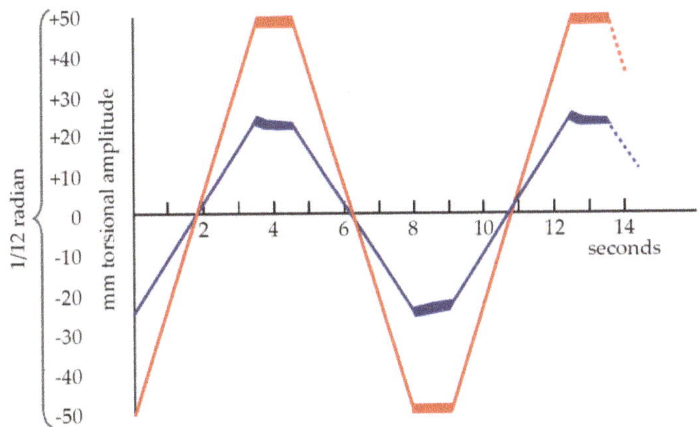

Figure 4. Chart representing the movement of Dr. Hartert's cup and pin after clotting of the blood sample. The red line represents the displacement of the cup and the blue line represents the displacement of the pin. The units on the x-axis represent the extent of the illuminated section of film by the rotating mirror. The units of mm amplitude on the y-axis remain today in many viscoelastic hemostatic assay systems. The units of mm to express a clot strength has been the source of confusion in this space. Reproduced with permission from [4].

Adoption of thromboelastography grew slowly following Dr. Hartert's initial work, limited primarily to research laboratories. It began to gain some momentum in the 1980s, particularly in high blood loss procedures such as liver transplantation [1] and cardiac surgery [7,8]. Two similar technologies were initially developed, Thromboelastography (Thrombelastograph® [TEG®] Hemostasis Analyzer) by Haemoscope, and Thromboelastometry (ROTEM®) by Tem International

GmbH. These two technologies defined the first 30 years of clinically adopted viscoelastic hemostatic assays. Both technologies were enhanced by the utilization of different assays to gain visibility on different aspects of clot formation. The assays and associated reagents have been developed to investigate different coagulation pathways, the contributions of platelets and fibrinogen to clot strength, and the effect of clot lysis.

As the use of viscoelastic hemostatic assays has grown into more applications, its clinical value has increased. This has attracted significant academic interest. As such, next-generation technologies are rapidly being developed. The remainder of this review will focus on the first-generation technologies that have pioneered this field, the next-generation devices that have more recently gained regulatory approval, and the experimental technologies on the horizon.

4. First-Generation Devices

4.1. TEG® 5000 (Haemoscope, Haemonetics)

The TEG® 5000 system was the first VHA device available, originally developed by Haemoscope and then acquired by Haemonetics. The system has a long clinical history and has been widely published in the literature [9–11]. The mechanism of the TEG® 5000 is very similar to what was originally developed by Dr. Hartert.

For TEG® 5000, the blood sample is placed in a heated cup and the cup is then oscillated through a 4°45′ rotation (Figure 5). A pin is suspended within the cup by a torsion wire. Before clotting, the shear that develops between the cup and pin results in a viscous shear of the test sample and therefore, no movement of the pin. As clot strength develops, the fluid gradually develops an elastic element, which results in movement of the pin. Instead of the light-based system used by Dr. Hartert, the TEG® 5000 uses an electromechanical proximity sensor to detect rotation of the pin. The total deflection of the pin is tracked and can be plotted as shown in Figure 6. A range of assays have been developed for the TEG® 5000 system, which includes those shown in Table 1.

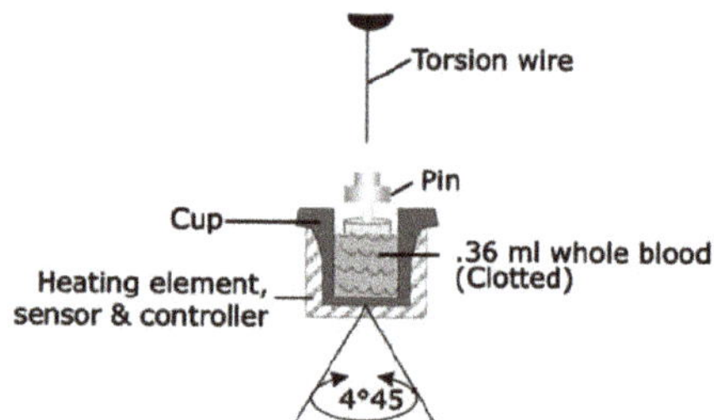

Figure 5. Schematic representation of the TEG® 5000 system. Thromboelastography conducted with the TEG® 5000 system uses approximately 0.36 mL of blood, which is placed into a cylindrical cup at 37 °C. A pin on a torsion wire is suspended in the blood, and the cup rotates in alternating directions (rotation angle 4°45′, cycle duration 10 s) to simulate venous flow. At the onset of each measurement, there is no torque between the cup and the pin, and the machine provides a reading of zero. As clotting occurs, fibrin fibers formed between the pin and the cup create a rotational force on the pin, which is measured via a torsion wire and an electromagnetic transducer; the readout line diverges from the baseline until it reaches a maximum value (maximum clot strength). With the onset of clot lysis, the readout converges back towards baseline. Reproduced with permission from Haemonetics [11].

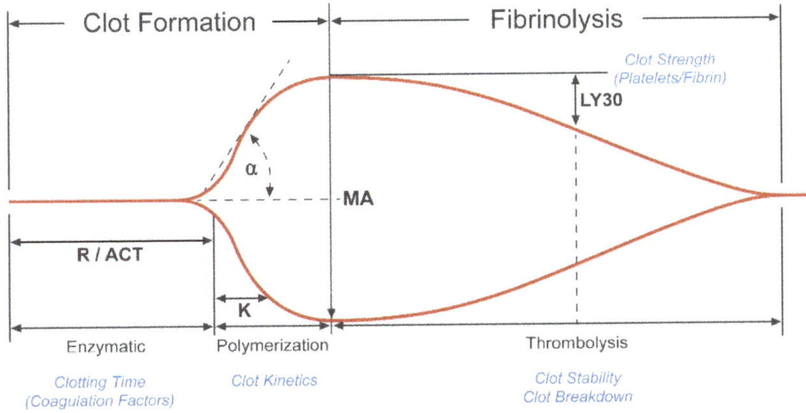

Figure 6. Plot of the total deflection of the pin from the TEG® 5000 system. The primary values that are derived from the resulting waveform are reaction time (R, or activated clotting time [ACT]), maximum amplitude (MA) and lysis at 30 minutes (LY30). R represents the time to the beginning of clot formation. MA is the maximum clot strength achieved. LY30 quantifies the reduction of clot strength, or the lysis (in the 30 min) following MA. K and α angle are also used to quantify the dynamics of clot formation. Reproduced from [12].

Table 1. Assays developed for the TEG® 5000 system.

TEG® Test	Description
Kaolin TEG®	Intrinsic pathway-activated assay.
Kaolin TEG® with Heparinase	Eliminates the effect of heparin. Used in conjunction with Kaolin TEG® to assess heparin effect.
RapidTEG™	Extrinsic and Intrinsic pathways to speed coagulation process and rapidly assess coagulation properties.
TEG® Functional Fibrinogen	Extrinsic pathway with platelet inhibitor to restrict platelet function. Allows for quantification of fibrinogen contribution to clot strength.
TEG® Platelet Mapping™	Utilises a platelet receptor-specific tracing in conjunction with Kaolin TEG®. Identifies level of platelet inhibition and aggregation.

4.2. *ROTEM® Delta (Tem International GmbH)*

The ROTEM® delta system was developed after the TEG® 5000 system and utilizes a similar approach with cups and pins making up the heart of the instrument. The difference lies in the actuation and detection of the cup and pin assembly. The cup is held stationary and heated, and a rotational force is applied to the pin. If there is no resistance to pin motion, the pin will rotate a full 4°75′. As the clot builds up, it restricts the rotation of the pin. The restriction of pin rotation is inversely proportional to the clot strength. Rotation of the pin is measured in a manner similar to that originally developed by Hartert; a collimated light beam from a light emitting diode is reflected off a mirror coupled to the pin. The motion of the pin is then tracked by a photosensor. As with the TEG® 5000, the ROTEM® delta system has a long clinical history. Various applications have been widely published [13,14] and a wide range of assays have been developed, including those shown below (Table 2).

Table 2. Assays developed for the ROTEM® delta system.

ROTEM® Test	Description
INTEM®	Contact activation with phospholipid and ellagic acid.
EXTEM®	Tissue factor activation.
HEPTEM®	Heparinase to neutralize heparin. Used in conjunction with INTEM® to assess heparin effect.
APTEM®	Inhibits fibrinolysis. Used in conjunction with EXTEM® to assess fibrinolysis.
FIBTEM®	Blocks platelet contribution to clot formation. Allows for quantification of fibrinogen contribution to clot strength.

4.3. *Sonoclot® (Sienco)*

Sonoclot® is another legacy device developed by Sienco. The Sonoclot® device differs from other legacy systems in that it is not a rotational-based system, but a linear motion system (Figure 7). A hollow, open-ended plastic probe is connected to the transducer head and immersed in a sample containing different coagulation activators or inhibitors, depending on the assay [15]. The probe oscillates vertically within the sample, with the oscillating motion modified as the blood sample begins to clot. The transducer outputs a signal that is proportional to the deflection of the test probe. The changes in the oscillatory pattern of the test probe are directly correlated to the viscoelastic properties of the sample. Measurements are taken over time and plotted along a time axis, similar to other legacy systems. The newest generation of the Sonoclot Analyzer is fully digitalized; however, it is not yet approved by the American Food and Drug Administration (FDA) and therefore is only utilized outside of the US.

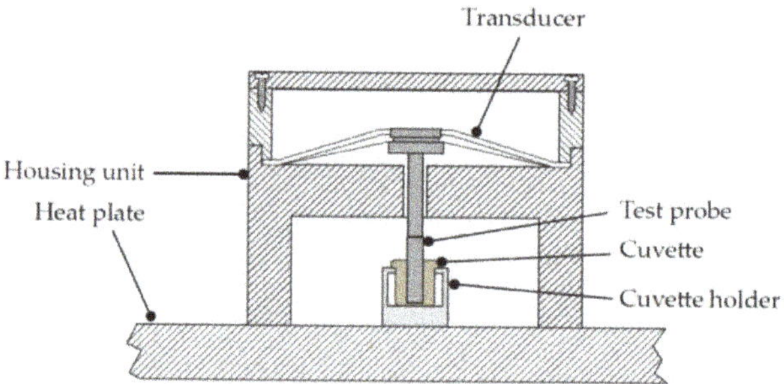

Figure 7. Schematic of the Sonoclot® system. The sample to be tested is placed in a cuvette within the housing unit. The linear test probe is then lowered into the cuvette and is in contact with the sample to be tested. An electric signal through the transducer produces a linear oscillatory motion of the test probe whilst a heat plate warms the sample to be tested through the cuvette holder. Adapted from US Patent 5,138,872 [16].

5. Towards Site of Care

The last two decades have seen rapid expansion of the use of viscoelastic hemostatic assays in an ever-expanding range of clinical applications. The devices described above did much to lay the groundwork for the clinical relevance of viscoelastic hemostatic assays. They offered a holistic view of the clotting process that simply was not available with traditional assay portfolios. However, the manual nature of performing the tests has limited the growth of its clinical utilization.

All of the first-generation devices require manual pipetting of blood samples and reagents. The open-nature of the systems limits their use in point-of-care applications. For many of the potential applications of this technology (cardiac surgery, trauma, etc.) speed and proximity to the patient are critical for success [17]. To overcome these limitations, the historic players in this market have developed next-generation systems. In addition, there are new entrants to this field with new technologies.

These next-generation technologies need to target several key factors to overcome the shortcomings of their predecessors.

- **Ease of use**—The ideal systems should decrease the overall number of manual transfer steps, or eliminate them altogether. Where possible, reagents should be dosed and reconstituted automatically. The user interface should be intuitive, with as little intervention as possible as the target users for these devices will extend beyond skilled medical laboratory technicians. Ideally, the systems would meet the requirements to be considered waived under Clinical Laboratory Improvement Amendments (CLIAs).
- **Size**—In order to optimize their use, the devices should be able to be utilized very close to the patient. This means they need to take up little valuable space in already crowded emergency rooms, cath labs and operating theatres. Taking up as little of this valuable real estate as possible is critical.
- **Resistance to ambient conditions**—Having a clear view of a patient's overall hemostasis as early as possible can be a lifesaving benefit in certain emergencies. In clinical scenarios such as stroke or massive trauma, getting feedback on the underlying hemostatic status of a patient as early as possible is critical. For this reason, next-generation devices may be used outside of traditional clinical settings. Environments such as ambulances, helicopters or even the battlefield should be considered. Fundamental technologies that may be sensitive to vibration or temperature may not be suitable for certain applications.
- **Remote viewing**—Fundamentally, these assays produce a sequence of results over a period of time and allow the entire clinical team to view results as they're being developed is important to timely intervention. Having the capability for readings to be displayed in real-time in the operating room may be critical. Allowing emergency department physicians to watch results develop whilst patients are being transported to the hospital will help prepare them for the arival of that patient.

Furthermore, unlike many of the first-generation devices that had significant inter-laboratory variance, with coefficients of variation >10% [18–20], these newer-generation devices have demonstrated reduced variance [21]. Whilst older-generation devices enabled users to focus on running channels of interest, newer devices detailed below require users to run entire cartridges. While this makes the assay process more streamlined and precise, it may be associated with increased cost.

5.1. TEG® 6s (Haemonetics)

To overcome the issues inherent with a manual device, Haemonetics released the TEG® 6s. The TEG® 6s is a cartridge-based system that automates all sample aliquoting, reagent mixing and testing. The system consists of a pneumatically controlled microfluidic cartridge that takes an initial blood sample and divides it across four distinct test channels. Each test channel includes a set of reagents to conduct four different assays simultaneously.

In order to bring the test methodology to a cartridge-based system, Haemonetics developed a new measurement technique. A fluid's shear modulus is proportional to its resonant frequency; the frequency at which a material will resonate when exposed to an excitation. The testing mechanism of the TEG® 6s consists of a droplet of fluid suspended between a light source and a photodetector (Figure 8). The droplet of fluid is excited by the displacement of a piezoelectric actuator. The piezoelectric actuator is driven with a function that includes all frequencies between 25 and 400 Hz, at increments of 0.25 Hz. When subjected to this range of frequencies, the sample will resonate at its resonant frequency. The optical sensors measure the displacement of the droplet and calculate the frequency of vibration with a Fourier transform.

The TEG® 6s system has gained regulatory clearance in the USA and Europe amongst other international markets and has been compared to first-generation thromboelastography systems in the literature [19,21–23]. The assays available with the TEG® 6s are similar to the TEG® 5000 and are available in three cartridge configurations. The Global Hemostasis cartridge includes Kaolin TEG® with Heparinase, RapidTEG™ and Functional Fibrinogen. It is also possible to have the Global Hemostasis cartridge without KaolinTEG® with Heparinase. The Platelet Mapping cartridge adds channels with adenosine diphosphate and arachidonic acid reagents to activate specific platelet activation pathways.

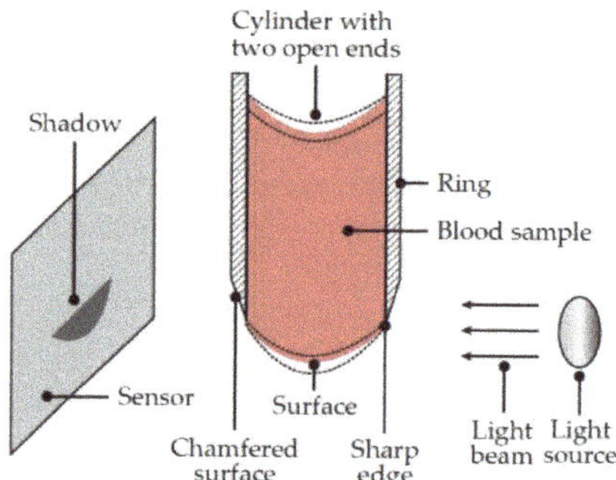

Figure 8. Schematic representation of the TEG® 6s. The diameter of the ring and cylinder is sized to allow the blood sample to be held within the cylinder by surface tension, without support at the bottom surface. The light source (light emitting diode) directs a beam of light at the blood sample, casting a shadow on the sensor. As the surface of the sample is excited by the light beam, the sample oscillates and the corresponding shadow becomes larger or smaller depending upon the elasticity of the blood sample. The resonant frequency of the blood sample is determined before, during and after coagulation. Changes in the resonant frequency of the sample are indicative of the hemostasis characteristics of the blood sample. Adapted from US Patent 7,879,615 B2 [24].

5.2. *ROTEM® Sigma (Instrumentation Laboratories)*

The ROTEM® sigma system is a transfer of the manual ROTEM® delta system to a cartridge-based system, which maintains the cup and pin methodology. The pins are embedded within the cartridge and actuated by the device when installed. The cartridge is capable of withdrawing a sample of blood from a tube and aliquoting it across four channels simultaneously. Fluid flow is controlled, and reagents are reconstituted automatically. The appropriate sample volume is deposited into each cup and the tests are initiated.

The ROTEM® sigma system offers two cartridges, the Sigma Complete and the Sigma Complete with Heparinase. Each cartridge provides the FIBTEM®, EXTEM®, INTEM® and APTEM® tests. The Heparinase cartridge allows for the neutralization of Heparin. The Sigma system has CE marking (Conformité Européene) for use in Europe, has been widely published, and been compared to the previous generation of thromboelastometry devices [23,25,26].

5.3. *Quantra® (HemoSonics)*

The Quantra® system developed by HemoSonics, which is CE marked and FDA cleared, utilizes Sonic Estimation of Elasticity via Resonance (SEER) technology [27]. This basic methodology has been used in the past in order to diagnose in vivo clots such as in deep vein thrombosis. Most diagnostic ultrasound devices use the echogenicity of a material in order to estimate its density. SEER technology analyzes the waveforms of the material and estimates the modulus of that material. By doing so, one can estimate if a clot has formed inside a vessel, and the extent of that clot. HemoSonics adapted this technology and refined it to get a more accurate estimate of the modulus of a fluid within a test chamber. The SEER technique has three distinct steps. The first step is data acquisition. This phase is broken up into a number of substeps. First, a low energy sensing pulse is sent into the test chamber. This sensing pulse provides a map of the location of acoustic scatterers within the sample. In the case of whole blood, the primary acoustic scatterers are red blood cells. Next, a higher energy forcing pulse is sent into the test chamber. This forcing pulse results in a displacement of the material within the test chamber. What follows is a series of sensing pulses to provide more maps of the acoustic scatterers. The next phase of the test is motion estimation. In order to estimate motion, a Finite Difference Time Domain model is utilized. This model outputs the relative motion of the acoustic scatterers. The output of this second phase is displacement vs. time waveform. The third phase of the test is to analyze the time displacement waveform for frequency. From this evaluation, the modulus of the fluid in the test chamber is estimated.

6. Emerging Technologies

Several emerging technologies are currently in development for point-of-care hemostatic testing, including microfluidics, fluorescent microscopy, electrochemical sensing, photoacoustic detection, and micro/nano electromechanical systems (MEMS/NEMS) [28]. In the section below, we outline several new devices/technologies that are currently under investigation for use as viscoelastic hemostatic assays; however, these technologies may also have other applications, such as monitoring microfluid dynamics and shear force in coagulation [29–34].

6.1. Laser Speckle Rheometry (Massachusettes General Hospital)

Laser Speckle Rheometry (Figure 9) is a technology being pioneered at the Wellman Center for Photomedicine at Massachusetts General Hospital and Harvard Medical School [35,36]. The technology has applications in a range of areas, including the catheter-based assessment of atherosclerotic plaques. One of the leading applications for the technology is as a viscoelastic hemostatic assay.

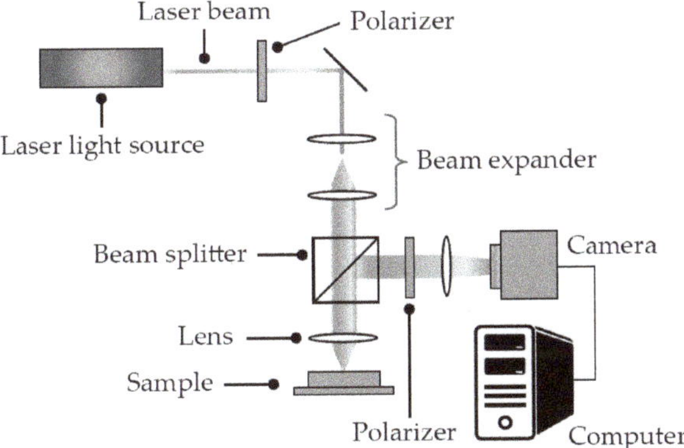

Figure 9. Schematic representation of the Laser Speckle Rheometry system. A laser light source shines a laser beam through a beam splitter onto a sample. The speckle pattern is detected by a camera and processed as described above. By analyzing the laser speckle pattern, the system can estimate the Brownian motion of the material. Brownian motion in such a sample is directly related to the viscoelastic properties of the material. During the blood clotting process, the change in viscoelastic properties can be evaluated and strongly correlated to parameters found in predicate viscoelastic hemostasis analyzers. Adapted from US Patent 8,772,039 B2 [35].

Laser speckle is a random intensity pattern that occurs by the interference of coherent light scattered from tissue. The technology uses a camera-based data acquisition system that detects the speckle pattern reflected from a material. A software algorithm detects changes in the speckle pattern over time. These changes are very sensitive to the passive Brownian motion of light scattering particles.

6.2. Mechanical Resonant Frequency (Abram Scientific)

Abram Scientific is developing a small, handheld approach to viscoelastic hemostatic monitoring [37]. The core technology is based on the concept that the resonant frequency of an object is affected by the medium that is surrounding that object. This can be described as viscoelastic damping. The measurement mechanism consists of a small vibrating element that is surrounded by the test sample. The vibrating element is actuated via an electric signal with a swept frequency. There is a detection mechanism that detects the resulting vibration frequency of the element. As the properties of the sample surrounding the vibrating element change, the frequency of vibration changes. In this way, the system is capable of detecting the viscoelastic properties of the sample. The Abrams Scientific device has the potential to be very compact with the vibrating element of the disposable device integrated into a layered microfluidic chip.

6.3. Ultrasonic Deformation (Levisonics)

Levisonics was formed with technology originally developed at Boston University and Tulane University [38]. This technology consists of an ultrasonic transducer that is positioned opposite a reflector. A standing ultrasonic wave is generated between the transducer and reflector. The radiation pressure generated is sufficient to balance the gravitational force on the sample, levitating it in position between a camera and light source. While the drop of blood is positioned at the pressure node of the standing wave, the amplitude can be modulated, which produces static or oscillatory shape deformations in the sample. These shape deformations are then recorded with a high-resolution digital camera or by laser scattering detection (Figure 10).

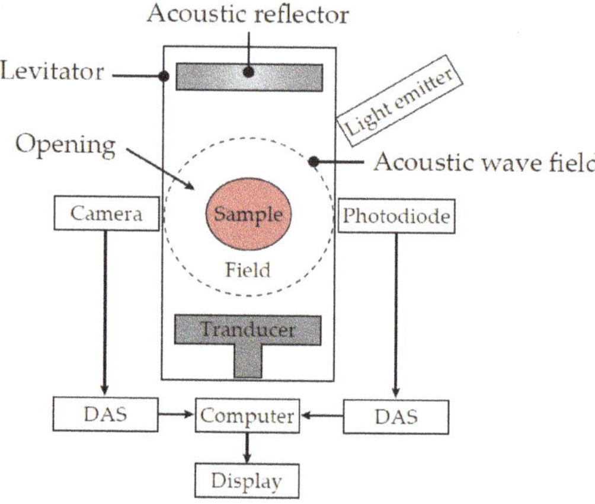

Figure 10. Schematic of the Levisonics ultrasonic transducer. The extent of shape deformations in the suspended blood droplet are directly correlated to the viscoelastic properties of the sample. As the amplitude of the standing wave is modulated, the shape of the sample will change from spherical to oblong. The camera system will track these shape changes and an algorithm converts them to viscoelastic properties. DAS = data acquisition system. Adapted from US Patent application 2017/0016878 A1 [38].

One potentially differentiating feature of this technology is the fact that the blood sample is able to be tested whilst not in contact with any artificial surfaces. This provides two potential benefits. First, there is no disruption to the fluid sample during the clotting process. This may eliminate any mechanical shear that could affect how the clot develops. Second, the absence of artificial surfaces being exposed to the sample during the clotting process may provide a more physiological environment.

6.4. Parallel Plate Viscometry (Entegrion)

Entegrion has developed a system that closely mimics traditional rheological test platforms used outside the medical industry [39]. In essence, they have developed a parallel plate viscometer (Figure 11). Their test cartridge consists of two parallel plates, each connected to a voice coil actuator. The sample to be tested lies between the two plates. Driving signals of various amplitudes and frequencies can elicit motion between the two plates. An optical system then detects the resulting motion on the sample. All the rheological information required to calculate the viscoelastic properties of the sample can be generated with this configuration. Entegrion has been able to integrate this technology into a small, point-of-care system. The resulting small footprint allows for applications in a range of clinical settings including pre-hospital emergency medicine.

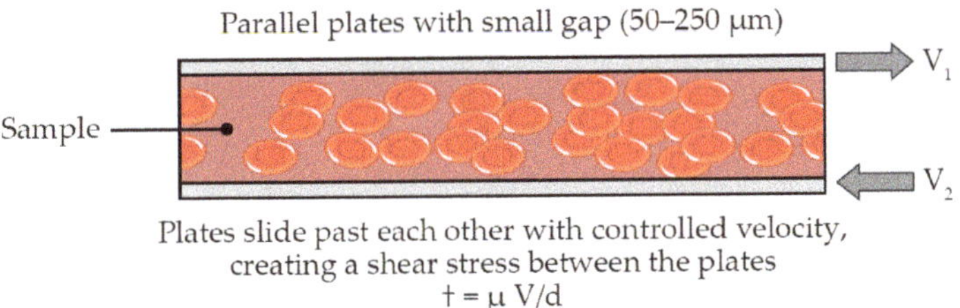

Figure 11. Parallel plate viscometry. Parallel plates slide past each other with controlled velocity to create a shear stress between the plates, which is represented as † = μV/d, where † = shear stress; μ = viscosity, $V = V_1 - V_2$, (relative linear velocity of the plates); d = gap between plates. Adapted from US Patent 8,450,078 B2 [39].

6.5. Traditional Viscoelastic Testing with "Active Tips" (Enicor)

Enicor has developed a novel solution for those who prefer the cost effectiveness and flexibility of an open system [40]. The measurement technology is very similar to the legacy devices such as TEG® and ROTEM®. With TEG®, the cup rotates and the resulting rotation on the pin is measured. With ROTEM®, the cup is held stationary and a rotational force is applied to the pin, which is then measured. With the ClotPro® system from enicor (Figure 12), the pin is held stationary and a rotational force is applied to the cup. The resulting rotation on the cup is measured.

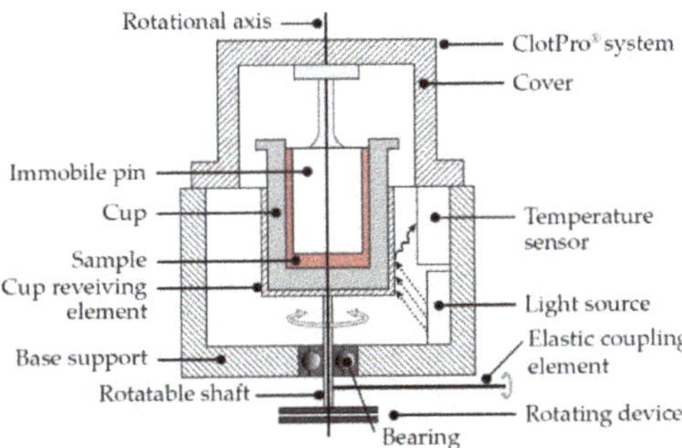

Figure 12. Schematic of the ClotPro® system. The ClotPro® system consists of a cup that contains the blood sample and a static pin, which is fixed to the cover. The cup rotates around the vertical axis, driven by an elastic coupling element, such as a spring wire, which is attached to the shaft. The light source (wavelength 1–3 μm) is placed within 75 mm of the shaft and cup-receiving element to act as a temperature control device. Adapted from International Patent WO,2018/137766,A1 [40].

With these legacy systems, multiple pipetting steps are required to meter and deliver the correct concentration of reagent into the sample cup. ClotPro® has simplified this by utilizing "Active Tips". Active tips are standard pipette tips that already contain the necessary amount of reagent. Blood is drawn into the tip and the reagent is automatically reconstituted in the correct dose. The blood is then dispensed into the cup and the test is initiated. The ClotPro® system also has a test menu similar to that of ROTEM®.

7. Conclusions

The field of viscoelastic hemostatic assays is rapidly growing and generating significant clinical studies, academic research, and industry funding. Whilst older devices have been available for several decades, new technologies are being introduced that promise to make viscoelastic testing more readily available in a wider range of clinical environments. New assays are being developed to give clinicians information about their patients' hemostasis that have never been available before. New indications that have been recently approved, or are currently under investigation, include trauma, interventional cardiology, extracorporeal cardiac assist, obstetrics and anti-coagulation therapy monitoring. The ability of viscoelastic hemostatic assays to deliver clear and readily actionable information about a patient's overall hemostatic status will continue to improve patient care in the years ahead; however, newer technologies and devices will have to demonstrate their clinical effectiveness.

Author Contributions: Initial literature search and review outline was generated by J.H. and M.M. M.M. drafted the initial manuscript with support from J.H. J.H., J.D.D. and M.M. reviewed the manuscript and provided substantial contributions to the final version. All authors have read and agreed to the published version of the manuscript.

Funding: This research received no external funding.

Acknowledgments: The authors thank Meridian HealthComms, Plumley, UK for providing formatting and editorial assistance in the submission process, which was funded by Haemonetics, SA, Signy, Switzerland, in accordance with Good Publication Practice (GPP3).

Conflicts of Interest: All authors are employees of Haemonetics, the manufacturer of the TEG® technology.

References

1. Kang, Y.G.; Martin, D.J.; Marquez, J.; Lewis, J.H.; Bontempo, F.A.; Shaw, B.W., Jr.; Starzl, T.E.; Winter, P.M. Intraoperative changes in blood coagulation and thrombelastographic monitoring in liver transplantation. *Anesth. Analg.* **1985**, *64*, 888–896. [CrossRef]
2. Mallett, S.V. Clinical utility of viscoelastic tests of coagulation (TEG/ROTEM) in patients with liver disease and during liver transplantation. *Semin. Thromb. Hemost.* **2015**, *41*, 527–537. [CrossRef]
3. Curry, N.S.; Davenport, R.; Pavord, S.; Mallett, S.V.; Kitchen, D.; Klein, A.A.; Maybury, H.; Collins, P.W.; Laffan, M. The use of viscoelastic haemostatic assays in the management of major bleeding: A British Society for Haematology Guideline. *Br. J. Haematol.* **2018**, *182*, 789–806. [CrossRef]
4. Hochleitner, G.; Sutor, K.; Levett, C.; Leyser, H.; Schlimp, C.J.; Solomon, C. Revisiting Hartert's 1962 Calculation of the Physical Constants of Thrombelastography. *Clin. Appl. Thromb. Hemost.* **2017**, *23*, 201–210. [CrossRef]
5. Hartert, H. Coagulation analysis with thromboelastography, a new method. *Klin. Wochenschr.* **1948**, *26*, 577–658. [CrossRef]
6. Hartert, H.; Schaeder, J.A. The physical and biological constants of thrombelastography. *Biorheology* **1962**, *1*, 31–39. [CrossRef]
7. Martin, P.; Horkay, F.; Rajah, S.M.; Walker, D.R. Monitoring of coagulation status using thrombelastography during paediatric open heart surgery. *Int. J. Clin. Monit. Comput.* **1991**, *8*, 183–187. [CrossRef]
8. Spiess, B.D.; Gillies, B.S.A.; Chandler, W.; Verrier, E. Changes in transfusion therapy and reexploration rate after institution of a blood management program in cardiac surgical patients. *J. Cardiothorac. Vasc. Anesth.* **1995**, *9*, 168–173. [CrossRef]

9. McNicol, P.L.; Liu, G.; Harley, I.D.; McCall, P.R.; Przybylowski, G.M.; Bowkett, J.; Angus, P.W.; Hardy, K.J.; Jones, R.M. Patterns of coagulopathy during liver transplantation: Experience with the first 75 cases using thrombelastography. *Anaesth. Intensive Care* **1994**, *22*, 659–665. [CrossRef]
10. Zuckerman, L.; Cohen, E.; Vagher, J.; Woodward, E.; Caprini, J. Comparison of thrombelastography with common coagulation tests. *Thromb. Haemost.* **1981**, *46*, 752–756. [CrossRef]
11. Hartmann, J.; Mason, D.; Achneck, H. Thromboelastography (TEG) point-of-care diagnostic for hemostasis management. *Point Care* **2018**, *17*, 15–22. [CrossRef]
12. Dias, J.D.; Norem, K.; Doorneweerd, D.D.; Thurer, R.L.; Popovsky, M.A.; Omert, L.A. Use of thromboelastography (teg) for detection of new oral anticoagulants. *Arch. Pathol. Lab. Med.* **2015**, *139*, 665–673. [CrossRef]
13. Entholzner, E.K.; Mielke, L.L.; Calatzis, A.N.; Feyh, J.; Hipp, R.; Hargasser, S.R. Coagulation effects of a recently developed hydroxyethyl starch (HES 130/0.4) compared to hydroxyethyl starches with higher molecular weight. *Acta Anaesthesiol. Scand.* **2000**, *44*, 1116–1121. [CrossRef]
14. Novakovic-Anucin, S.; Kosanovic, D.; Gnip, S.; Canak, V.; Cabarkapa, V.; Mitic, G. Comparison of standard coagulation tests and rotational thromboelastometry for hemostatic system monitoring during orthotopic liver transplantation: Results from a pilot study. *Med. Pregl.* **2015**, *68*, 301–307. [CrossRef]
15. Ganter, M.T.; Hofer, C.K. Coagulation monitoring: Current techniques and clinical use of viscoelastic point-of-care coagulation devices. *Anesth. Analg.* **2008**, *106*, 1366–1375. [CrossRef]
16. Henderson, J.H. Fluid Viscoelastic Test Apparatus and Method. US Patent 5,138,872, 18 August 1992.
17. Evans, A.; Sutton, K.; Hernandez, S.; Tanyeri, M. Viscoelastic Hemostatic Assays - A Quest for Holy Grail of Coagulation Monitoring in Trauma Care. *Bioeng. Ann. J.* **2019**, *1*, 61–64.
18. Chitlur, M.; Sorensen, B.; Rivard, G.E.; Young, G.; Ingerslev, J.; Othman, M.; Nugent, D.; Kenet, G.; Escobar, M.; Lusher, J. Standardization of thromboelastography: A report from the TEG-ROTEM working group. *Haemophilia* **2011**, *17*, 532–537. [CrossRef]
19. Gurbel, P.A.; Bliden, K.P.; Tantry, U.S.; Monroe, A.L.; Muresan, A.A.; Brunner, N.E.; Lopez-Espina, C.G.; Delmenico, P.R.; Cohen, E.; Raviv, G.; et al. First report of the point-of-care TEG: A technical validation study of the TEG-6S system. *Platelets* **2016**, *27*, 642–649. [CrossRef]
20. Dias, J.D.; Pottgiesser, T.; Hartmann, J.; Duerschmied, D.; Bode, C.; Achneck, H.E. Comparison of three common whole blood platelet function tests for in vitro P2Y12 induced platelet inhibition. *J. Thromb. Thrombolysis* **2019**. [CrossRef]
21. Neal, M.D.; Moore, E.E.; Walsh, M.; Thomas, S.; Callcut, R.A.; Kornblith, L.Z.; Schreiber, M.; Ekeh, A.P.; Singer, A.J.; Lottenberg, L.; et al. A comparison between the TEG 6s and TEG 5000 analyzers to assess coagulation in trauma patients. *J. Trauma Acute Care Surg.* **2020**, *88*, 279–285. [CrossRef]
22. McCormack, D.; El-Gamel, A.; Gimpel, D.; Mak, J.; Robinson, S. A 'real-world' comparison of TEG® 5000 and TEG® 6s measurements and guided clinical management. *Heart Lung Circ.* **2019**, *28*, S79. [CrossRef]
23. Ziegler, B.; Voelckel, W.; Zipperle, J.; Grottke, O.; Schöchl, H. Comparison between the new fully automated viscoelastic coagulation analysers TEG® 6s and ROTEM® *sigma* in trauma patients: A prospective observational study. *Eur. J. Anaesthesiol.* **2019**, *36*, 834–842. [CrossRef] [PubMed]
24. Kautzky, H. Hemostasis Analyzer and Method. US Patent 7,879,615 B2, 1 February 2011.
25. Gillissen, A.; van den Akker, T.; Caram-Deelder, C.; Henriquez, D.D.C.A.; Bloemenkamp, K.W.M.; Eikenboom, J.; van der Bom, J.G.; de Maat, M.P.M. Comparison of thromboelastometry by ROTEM® *delta* and ROTEM® *sigma* in women with postpartum haemorrhage. *Scand. J. Clin. Lab. Investig.* **2019**, *79*, 32–38. [CrossRef]
26. Schenk, B.; Görlinger, K.; Treml, B.; Tauber, H.; Fries, D.; Niederwanger, C.; Oswald, E.; Bachler, M. A comparison of the new ROTEM® *sigma* with its predecessor, the ROTEM® *delta*. *Anaesthesia* **2019**, *74*, 348–356. [CrossRef]
27. Viola, F.; Kramer, M.; Walker, W.; Lawrence, M. Method and Apparatus for Characterization of Clot Formation. US Patent 7,892,188B2, 22 February 2011.
28. Mohammadi Aria, M.; Erten, A.; Yalcin, O. Technology Advancements in Blood Coagulation Measurements for Point-of-Care Diagnostic Testing. *Front. Bioeng. Biotechnol.* **2019**, *7*, 395. [CrossRef]
29. Colace, T.V.; Fogarty, P.F.; Panckeri, K.A.; Li, R.; Diamond, S.L. Microfluidic assay of hemophilic blood clotting: Distinct deficits in platelet and fibrin deposition at low factor levels. *J. Thromb. Haemost.* **2014**, *12*, 147–158. [CrossRef]

30. Colace, T.V.; Tormoen, G.W.; McCarty, O.J.; Diamond, S.L. Microfluidics and coagulation biology. *Annu. Rev. Biomed. Eng.* **2013**, *15*, 283–303. [CrossRef]
31. Jain, A.; Graveline, A.; Waterhouse, A.; Vernet, A.; Flaumenhaft, R.; Ingber, D.E. A shear gradient-activated microfluidic device for automated monitoring of whole blood haemostasis and platelet function. *Nat. Commun.* **2016**, *7*, 10176. [CrossRef]
32. Judith, R.M.; Fisher, J.K.; Spero, R.C.; Fiser, B.L.; Turner, A.; Oberhardt, B.; Taylor, R.M.; Falvo, M.R.; Superfine, R. Micro-elastometry on whole blood clots using actuated surface-attached posts (ASAPs). *Lab. Chip.* **2015**, *15*, 1385–1393. [CrossRef]
33. Maji, D.; De La Fuente, M.; Kucukal, E.; Sekhon, U.D.S.; Schmaier, A.H.; Sen Gupta, A.; Gurkan, U.A.; Nieman, M.T.; Stavrou, E.X.; Mohseni, P.; et al. Assessment of whole blood coagulation with a microfluidic dielectric sensor. *J. Thromb. Haemost.* **2018**, *16*, 2050–2056. [CrossRef]
34. Ting, L.H.; Feghhi, S.; Taparia, N.; Smith, A.O.; Karchin, A.; Lim, E.; John, A.S.; Wang, X.; Rue, T.; White, N.J.; et al. Contractile forces in platelet aggregates under microfluidic shear gradients reflect platelet inhibition and bleeding risk. *Nat. Commun.* **2019**, *10*, 1204. [CrossRef] [PubMed]
35. Nadkarni, S. Optical Thromboelastography System and Method for Evaluation of Blood Coagulation Metrics. U.S. Patent 8,772,039B2, 8 July 2014.
36. Nadkarni, S. Comprehensive coagulation profiling at the point-of-care using a novel laser-based approach. *Semin. Thromb. Hemost.* **2019**, *45*, 264–274. [CrossRef] [PubMed]
37. Abhishek, R.; Yoshimizu, N. Methods, Devices, and Systems for Measuring Physical Properties of Fluid. U.S. Patent 9,909,968B2, 6 March 2018.
38. Trustees of Boston University. Apparatus, Systems & Methods for Non-Contact Rheological Measurements of Biological Materials. U.S. Patent Application 2017/0016878A1, 19 January 2017.
39. Dennis, R.; Fischer, T.; DaCorta, J. Portable Coagulation Monitoring Device and Method of Assessing Coagulation Response. U.S. Patent 8,450,078B2, 28 May 2013.
40. Dynabyte Informationssysteme GmbH. Devices and Methods for Measuring Viscoelastic Changes of a Sample. International Patent WO 2018/137766,A1, 2 August 2018.

Review

Role of pH Value in Clinically Relevant Diagnosis

Shu-Hua Kuo [1], Ching-Ju Shen [2], Ching-Fen Shen [3,*] and Chao-Min Cheng [1,*]

1 Institute of Biomedical Engineering, National Tsing Hua University, Hsinchu 300, Taiwan; mitsunari0408@gmail.com
2 College of Medicine, Kaohsiung Medical University, Kaohsiung 807, Taiwan; chenmed.tw@yahoo.com.tw
3 Department of Pediatrics, National Cheng Kung University Hospital, College of Medicine, National Cheng Kung University, Tainan 704, Taiwan
* Correspondence: drshen1112@gmail.com (C.-F.S.); chaomin@mx.nthu.edu.tw (C.-M.C.)

Received: 17 December 2019; Accepted: 14 February 2020; Published: 16 February 2020

Abstract: As a highly influential physiological factor, pH may be leveraged as a tool to diagnose physiological state. It may be especially suitable for diagnosing and assessing skin structure and wound status. Multiple innovative and elegant smart wound dressings combined with either pH sensors or drug control-released carriers have been extensively studied. Increasing our understanding of the role of pH value in clinically relevant diagnostics should assist clinicians and improve personal health management in the home. In this review, we summarized a number of articles and discussed the role of pH on the skin surface as well as the factors that influence skin pH and pH-relevant skin diseases, but also the relationship of skin pH to the wound healing process, including its influence on the activity of proteases, bacterial enterotoxin, and some antibacterial agents. A great number of papers discussing physiological pH value have been published in recent decades, far too many to be included in this review. Here, we have focused on the impact of pH on wounds and skin with an emphasis on clinically relevant diagnosis toward effective treatment. We have also summarized the differences in skin structure and wound care between adults and infants, noting that infants have fragile skin and poor skin barriers, which makes them more vulnerable to skin damage and compels particular care, especially for wounds.

Keywords: pH value; diagnosis; skin; wound

---

1. Introduction

Skin, the major portion of the integumentary system, is the human body's largest organ. It spans approximately 2 m^2 and is 0.075–0.15 mm thick in the average adult [1]. It acts as the first line of defense to shield the human body from ultraviolet radiation, infection by pathogens, and chemical irritants [2]. Furthermore, skin plays a key role in thermoregulation, immunological function, and maintaining body water balance.

Schade and Marchionini first introduced the term "acid mantle" to describe the slight acidification of the uppermost layer of human skin [3]. The acid mantle is characterized by a pH value of 4–6, as a result of amino acids, fatty acids, sebum secreted by the sebaceous gland, and lactate excreted from sweat. All of these compounds are acidic and when present together on the human skin, become a barrier that prevents bacterial colonization. Another critical skin barrier is the lipid barrier, the extracellular lipid matrix of the stratum corneum (SC) [4], which is composed of free fatty acids, cholesterol and ceramides and functions as a hydrophobic barrier for human skin [5]. A skin barrier schematic is provided in Figure 1.

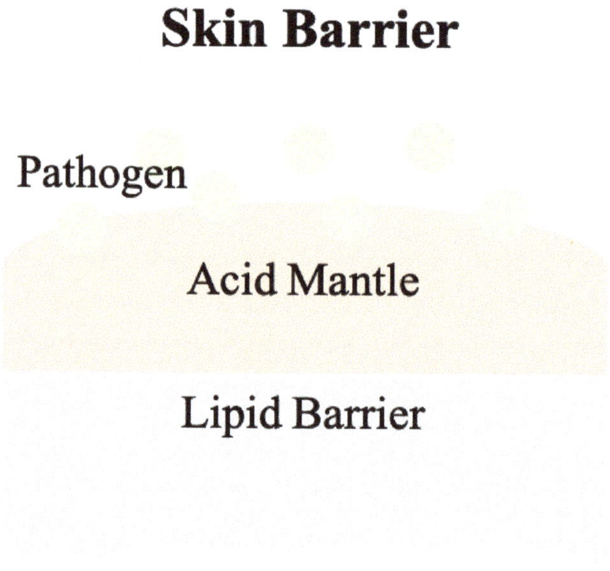

Skin Barrier

Pathogen

Acid Mantle

Lipid Barrier

Figure 1. The scheme of the skin barrier. Lipid barrier sits atop the acid mantle and displays a pH value of 4–6. Pathogens can be various bacteria. Slight acidic environment is a disadvantage for bacterial colonization.

Most human-pathogen bacteria are inhibited by the acidic milieu on the surface of normal human skin, but this shield is disturbed when skin is wounded. The tissue beneath skin has a physiological pH of 7.4 [6], which raises the overall pH value at the wound site and provides advantages for bacterial colonization. Bacterial contaminants in wounds, e.g., *Staphylococcus aureus*, interfere with the normal wound healing process.

There are four phases in normal wound healing [7,8]: hemostasis, inflammation, proliferation, and remodeling. During hemostasis the wound is filled with fibrin and coagulated blood. Clots are formed to stop bleeding and seal the wound site until tissues are repaired. During inflammation, histamine is produced by basophils and mast cells to increase capillary permeability, which allows leukocytes such as neutrophils to migrate to the infected wound site and remove dead cells and pathogens. However, a failure in eliminating pathogens or the remaining high pH environment can be primary reasons for chronic wound development. During proliferation, epithelial cells increase in number, granulation tissue forms, and angiogenesis occurs. Healthy granulation tissues should be pink in color and uneven in texture. The final phase, remodeling, may take years to complete. The wound matrix undergoes degradation by metalloproteinases and new extracellular matrix (ECM) is created during remodeling. Scar tissue can form due to degradation or ECM generation disruption. While the alkaline milieu activates protease, which facilitates the removal of damaged components, excessive amounts of protease eventually destroy newly constructed tissue [9]. Moreover, alkaline pH values are usually found in chronic wounds, and are associated with an increased risk of bacterial colonization [10].

pH value is a critical factor in the wound healing process and could offer important physiological condition information regarding skin status and infection. It assists physicians with clinically relevant diagnosis and patient wounds care. It can be used, for instance, to determine whether wounds are infected, becoming chronic, provide monitored, real-time wound status, and ultimately facilitate proper treatment in a more timely manner.

Recently, multiple innovative and elegant smart wound dressings combined with either pH sensors or drug control-released carriers have been extensively studied [11–15]. With these promising technologies, wound monitoring procedures can be visualized and designed for use in clinics and at home. The first step in clinically managing and in self-managing wounds is to understand wound status. Advances in and implementation of such technologies may alter the course of wound care and improve quality of life.

The role of pH value in evaluating skin and wound status will be systematically discussed in this review. We will emphasize its clinical application and highlight significant research that leverages pH to promote precise diagnoses leading to the development of impactful medical devices. We will also discuss pH-related differences in caring for adult skin versus neonatal skin, which is thin and requires more rigorous efforts for effective care.

2. Skin

As the vital, first line of defense between the body and the environment, skin is indispensable to human life. Skin structure has been thoroughly described in the literature. It is made up of three major strata, the epidermis, the dermis, and subcutaneous tissue. The epidermis, the outermost layer in skin structure, is approximately 0.05–1 mm thick, provides barrier functions, prevents infection, and regulates transepidermal water loss (TEWL). This layer is constructed of multiple sub-layers; stratum corneum, stratum lucidum, stratum granulosum, stratum spinosum, and stratum basale. The acid mantle and lipid barrier can be found on top of the epidermis, specifically just above the stratum corneum. Figure 2a is a schematic illustrating the structure of the epidermis. The second of the skin's major strata, the dermis, can be divided into two layers, the papillary region and the reticular region. The papillary region is the thin and superficial area adjacent to the epidermis. The subsequent layer, the reticular region, is thick and contains arector pili muscles, sebaceous glands, sweat gland ducts, merocrine sweat glands, hair follicles, blood capillaries, sensory receptors, and nerve fibers. The dermis layer is approximately 1–2 mm thick, and comprises collagen, elastic fibers, and extrafibrillar matrix for providing support. The third and deepest of skin's major three strata, subcutaneous tissue, contains fibroblasts, adipose cells, and macrophages. Subcutaneous tissue is also known as hypodermis and ranges in thickness from 1.65 mm to 18.20 mm depending on gender and skin site [16]. The hypodermis layer is used for fat storage, but it also contains large quantities of loose connective tissue and blood vessels. The structure of the dermis and hypodermis layers are depicted in Figure 2b.

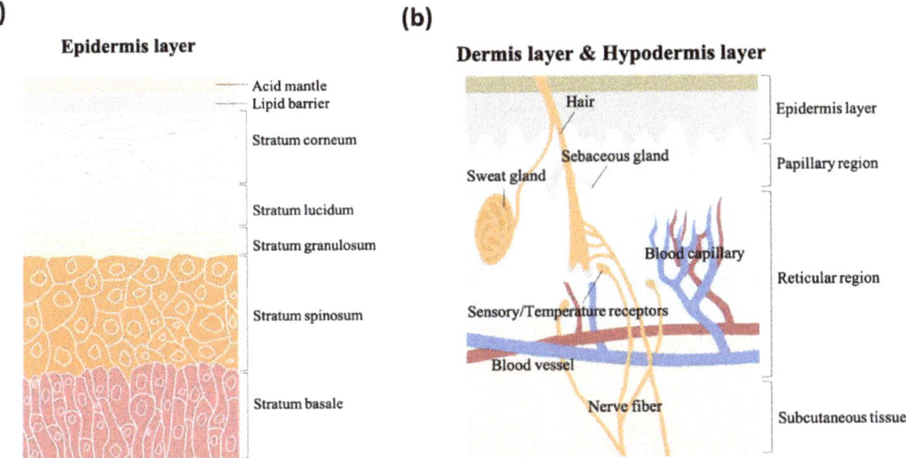

Figure 2. The schematic structure of (**a**) epidermis layer, (**b**) dermis layer, and hypodermis layer.

2.1. pH Values on Skin Surfaces

The skin pH of newborns is weakly alkaline (pH of 7.4) as a result of being encapsulated within the amniotic fluid and the vernix caseosa. Hans, et al., determined and described skin pH characteristics from 209 newborn infants [17]. However, after birth, the weakly alkaline skin becomes acidic in its quest to form the acid mantle [18,19] and the functional maturation of stratum corneum is accelerated within 2 to 8 weeks [20]. During this period, infant skin remains highly prone to damage.

During sexual maturation, the pH of skin in the axillary vault climbs from around 5.0 to near 7.0, and then returns to a lower pH during sexual involution [21,22]. A. Zlotogorski [23] measured the pH distribution on the surface of the skin (forehead and cheek) of 574 men and women aged 18 to 95, and found that those over the age of 80 had higher skin surface pH values compared to the other groups, which displayed skin surface pH values between 4.0 and 5.5 on the forehead and between 4.2 and 5.9 on the cheek. There was no significant pH difference based on gender. Irvin H. Blank [24] obtained skin surface pH values (forearm, antecubital, elbow, upper arm, forehead, and back of neck) from 100 males and 100 females aged 19–27 and found pH values varying from 4.0 to 7.0, with the most frequent readings being between 4.2 and 5.6, and evidence suggesting that females have a slightly higher skin pH (0.5 higher) than males. C. Ehlers and coworkers [25] found that there was a statistically significant difference in skin pH (arm) between men and woman. Notably, measurements could not be conducted close to the wrist because most people wash their hands several times each day, and soaps can make the skin more alkaline. Another study conducted by S. Luebberding and coworkers [26] followed strict criteria and found pH differences between males and females in TEWL, SC hydration, sebum content, and at the skin surface (forehead, forearm, hand and cheek). Table 1 (adapted from [26]) displays gender-related data showing that the pH values from five skin sites and across all age groups were higher in females compared to males. This data also shows that the mean evaporimetry and corneometry values were higher for the females of this group, while the mean sebumetry values were higher for the males.

It is common to obtain inconsistent pH measurements from men and women, but additional research could provide data to support a pattern of pH value differences between genders.

2.2. Factors That Affects Skin pH

Regulating skin surface pH is critical to physiological defense mechanisms, but skin is affected by a number of endogenous and exogenous factors [27], including age, ethnicity, sebum, sweat, detergents, cosmetic products, soaps, occlusive dressing, and more. These factors and their influence are displayed in Table 2 [23,26–47].

Elizabeth A. and coworkers [48] analyzed the distribution of the human skin microbiome via examination of 16S ribosomal RNA gene sequences, and found that bacteria in sebaceous sites where *Propionibacterium acnes* predominated were less diverse, less even, and less rich than the bacterial milieu in moist and dry sites, which were predominantly associated with Firmicutes and Proteobacteria. Multiple factors such as sweat, sebum secretion, and pH could have correlation with predominant microbiota. This study revealed a pattern of bacterial distribution on human skin that may have implications for the future treatment of pH-relevant skin disorders, a topic that we will explore further.

Table 1. Mean values (MV ± SD) arranged by age groups and localization (adapted from [26]).

	AG I		AG II		AG III		AG IV		AG V		Mean	
	♀	♂	♀	♂	♀	♂	♀	♂	♀	♂	♀	♂
F	5.01 ± 0.42	4.31 ± 0.28	4.98 ± 0.50	4.47 ± 0.33	4.92 ± 0.40	4.50 ± 0.36	5.10 ± 0.49	4.56 ± 0.30	4.67 ± 0.32	4.57 ± 0.35	4.94 ± 0.45	4.48 ± 0.33
C	5.33 ± 0.32	4.66 ± 0.38	5.27 ± 0.44	4.77 ± 0.33	5.19 ± 0.29	4.74 ± 0.36	5.35 ± 0.46	4.82 ± 0.27	5.21 ± 0.41	4.93 ± 0.32	5.27 ± 0.39	4.78 ± 0.34
N	5.09 ± 0.32	4.62 ± 0.37	5.09 ± 0.36	4.72 ± 0.36	5.08 ± 0.34	4.63 ± 0.34	5.37 ± 0.49	4.76 ± 0.28	5.05 ± 0.41	4.73 ± 0.29	5.13 ± 0.40	4.69 ± 0.33
A	5.12 ± 0.31	4.49 ± 0.42	5.22 ± 0.40	4.50 ± 0.36	5.17 ± 0.38	4.53 ± 0.30	5.44 ± 0.45	4.62 ± 0.37	5.12 ± 0.43	4.58 ± 0.37	5.21 ± 0.41	4.54 ± 0.36
H	5.10 ± 0.37	4.33 ± 0.36	5.19 ± 0.40	4.41 ± 0.32	5.03 ± 0.35	4.38 ± 0.31	5.25 ± 0.38	4.55 ± 0.42	4.88 ± 0.41	4.46 ± 0.40	5.09 ± 0.40	4.42 ± 0.37

	Female (Age ± SD)	Male (Age ± SD)
AG I	24.73 ± 2.59	25.70 ± 2.48
AG II	33.30 ± 2.92	33.23 ± 2.90
AG III	43.63 ± 2.81	44.23 ± 3.09
AG IV	55.17 ± 2.84	54.20 ± 3.13
AG V	66.57 ± 4.31	66.63 ± 2.61

F = Forehead; C = Cheek; N = Neck; A = Forearm; H = Hand.

Table 2. Factors that have influences on skin pH.

Factors	Influences	References
Age	Increased age is associated with skin pH values that are in steady fluctuation. This contributes to difficulties in making comparisons. Females in [26] show significantly higher skin pH values than males.	[23,26,41]
Ethnic	Whites have slightly higher pH than blacks, but there are mostly no statistically significant differences between ethnic groups.	[42–44]
Sebum	There is a negative correlation between sebum secretion and skin pH that is statistically significant in only the skin U-zone with acne and in the skin T-zone without acne, yet the correlation is not strong enough to explain the impact of sebum secretion on skin pH.	[45]
Sweat	Sweat is predominantly composed of sodium chloride, lactic acid, urea, and fatty acid. The acidic nature of human skin is partially due to lactic acid in sweat. As the rate of eccrine sweat increases, the concentrations of ammonia, pyruvic acid and lactate decrease, hence the pH increases.	[28–30,46,47]
Soaps and detergents	The use of soaps, synthetic detergents and even tap water will raise skin pH in a short-term effect.	[27,31–35]
Cosmetic products	Skin pH values drop after the application of cosmetic products, which can buffer the impact of tap water washing.	[36,37]
Occlusive dressings	Under occlusion, skin pH increases, but after removing occlusive dressings, the pH milieu drops back to an acidic nature. The application of occlusion has a short period effect on skin pH values.	[38–40]

2.3. Skin Diseases and pH Values

Beneath the acid mantle and lipid barrier, the various strata of skin have relatively static and regulated physiological pH values, but when disruptions in the protective barriers occur, or when normal skin pH is disrupted, the risk for specific skin diseases increases. Many studies have suggested a relationship between skin surface pH and disease.

It has been reported that great numbers of *Staphylococcus aureus* exist on the skin of patients with eczema can exacerbate the disease [49]. *Staphylococcus aureus* has a capacity for adhering to atopic corneocytes; this adherence phenomenon is partially attributable to the presence of protein A in its wall [50]. The pH range for optimal growth of *Staphylococcus aureus* is between 7.0 and 7.5, yet the bacteria are able to reproduce at pH values from 4.5 to 9.0. The pH value of the normal skin surface, 4.5–6.0, is consequently insufficient to kill the bacteria completely. However, the staphylococcal enterotoxin C2 (SEC2) can be destroyed in strongly acidic pH. Zinc ion, present in the crystalline structure of SEC2, may be removed completely in this process, which distorts SEC2 from its normal configuration [51].

pH on the skin surface increases in patients with seborrheic dermatitis [52–54]. *Pityrosporum ovule* prosper in scale of patients with dandruff, since higher pH values also promote the growth of yeast. Its colonization in less acidic areas of the skin result in declining activity of calf thymus histones, parotid saliva, lysozyme, and proteins obtained from human neutrophil granulocyte, which are cationic substances critically related to bacterial activity [55–58].

Diaper dermatitis is frequently diagnosed in infants, and wearing diapers is associated with increases in wetness and skin pH value. Ronald and coworkers [59] carried out a study involving 1601 infants. Their results suggest that diapered skin has approximately twice the TEWL of undiapered skin. Diapered skin has a higher pH value (5.9) while undiapered skin (pH 5.3) remains related to the acid mantle pH. Skin wetness could disturb the integrity of skin by increasing its permeability to irritants, frictional coefficient, and promoting microbial growth [60]. Elevated pH can increase the activity of fecal proteolytic enzymes that can degrade the extracellular matrix of the skin and eventually destroy tissues [61].

Skin pH is believed to be one of possible factors that promote candida infection, which diabetics are particularly prone to suffer from. Dimorphism of *Candida albicans* exists, as the blastospore form is related to acidic pH and the mycelial form is associated with alkaline pH [62,63]. The mycelial form of *Candida albicans* has been verified in clinical studies via its pathogenicity [64–66]. Gil Yosopovitch and coworkers [67] found that diabetic patients had significantly higher skin pH than that of normal control skin. Moreover, the skin pH values of diabetic patients with candidiasis was higher than that of diabetic patients without candidiasis. This suggests a possible correlation between skin pH and candida infection.

2.4. Skin Care

There are several structural differences between the skin of newborns and that of adults. Because newborn skin has a poorly functional skin barrier, and because newborns have fragile skin that can be easily damaged, proper and adequate steps must be taken to provide protection. S. Dhar [68] suggested that newborn bathing should not last longer than 5 min, or it could increase the hydration of the skin and decrease the friction threshold. Moreover, soaps and cleansers should be minimally used in the first few weeks of life for newborns, because any attempt intended to raise the skin pH would promote the number of bacteria and increase the TEWL [69]. There is not enough evidence that soaps have long-term impact on infants, but in the short term, it could disturb their acid mantles [31,33,34]. Synthetic detergents such as cocoyl isethionate, and sodium lauryl sulphate are soap substitutes that have pH values similar to skin and are more moderate than soaps. They do not alter skin pH and microflora, but they disintegrate rapidly and can cause skin dryness [70].

The spread of lesions in infantile seborrheic dermatitis can be limited by mineral oil or vegetable oil and minimizing the use of shampoos for cleaning the hair. Sarkar et al. [71] recommended the use of coconut oil as an emollient for application to neonatal skin due to its easy availability and economy. Skin irritation can also be avoided by using fragrance-free baby products.

For adult skin care, Y.C. Jung and coworkers [72] suggested that low skin pH could be maintained by increasing hydration, the presence of low skin pH induced higher sebum excretion rate, and the combined effects could suppress skin aging by reducing wrinkle length and depth. In work by S. H. Youn et al. [73], they found it helpful to control acne in the T-zone by increasing skin pH to 5.5–6 for female acne patients, and by decreasing skin pH to approximately 5.5 for male patients. Saba et al. [74] recommended that the ideal pH for body wash, soap, or cleanser would be in the range of 4.5–6.5, which is similar to that of the skin's acid mantle. Moreover, syndets, synthetic detergents, are less irritating and preferred. Nix and coworkers [75] suggested that the selection of skin cleaning products should take multiple factors into consideration, namely formulation, ingredients, skin compatibility, pH, and related infection control issues. Skin cleaning products with a pH value higher than 7 can disrupt skin barrier function. It is particularly important for elderly patients to select products with a pH value of 4–7 because the skin of the elderly is dryer, more prone to cracking, and recovers more slowly from damage caused by alkalinity.

3. Wounds

Wounds are divided into two categories, acute and chronic. Acute wounds can be healed predictably with normal wound healing approaches, and chronic wounds develop following one or more failures in the normal wound healing process. The wound healing process has four phases and has been described in detail above.

Acute wounds, sudden injuries to the skin, vary from superficial scratches to deep skin damages that can happen anywhere on the body. The causes of acute wounds vary but primarily include abrasion, puncture, laceration, and incision. Classified by causality, the two dominant types of acute wounds are surgical and traumatic. Surgical wounds are intentionally created for medical reasons, and traumatic wounds are randomly caused by external force. Severe pain is associated with wounds, which are especially prevalent among patients with fragile skin.

Chronic wounds remain in an inflamed state that delays healing for long periods lasting several months. Chronic wounds may take years to recover or may never heal, leading to physical and psychological suffering as well as considerable pressure on the social healthcare system. Diabetic patients with chronic wounds are at especially high risk for bacterial infections, because the slightly alkaline pH of chronic wounds promotes bacterial colonization. Figure 3a shows the pH values of acute wounds and chronic wounds in Figure 3b (adapted from [10]). During healing, the pH of acute wounds gradually declines to a level approximating that of the acid mantle. Chronic wounds, however, remain slightly alkaline. Monitoring and manipulating pH can be a critical tool for preventing and treating chronic wounds. Studies have reported on pH value monitoring combined with pH-activated drug control release systems and smart wound dressings and bandages [11–15].

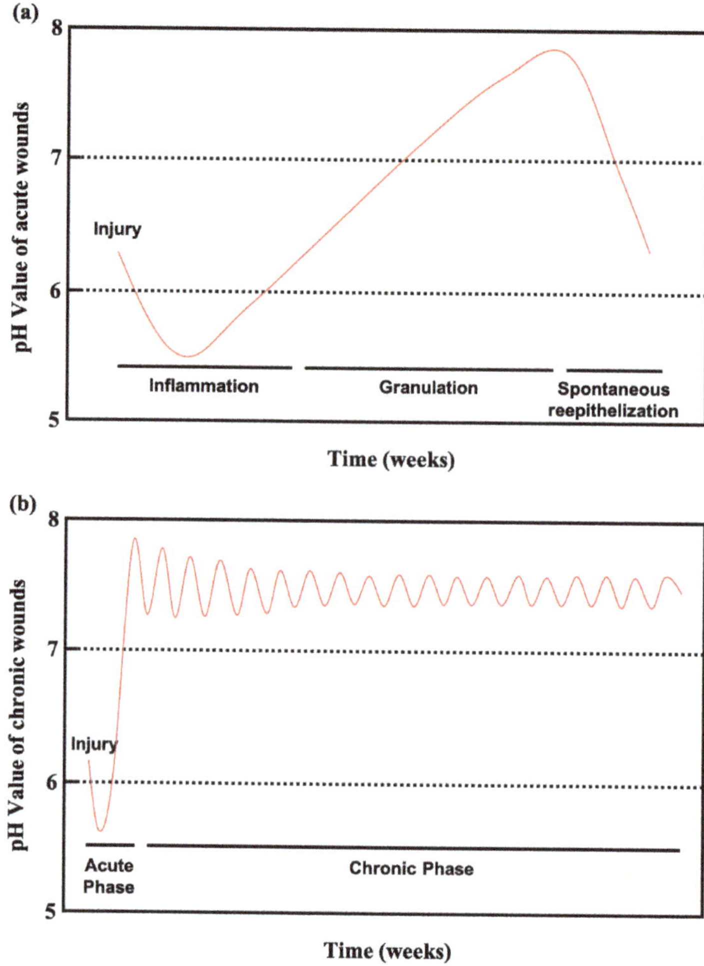

Figure 3. (**a**) Course of the pH milieu in acute wounds. (**b**) Course of the pH milieu in chronic wounds (adapted from [10]).

Severe chronic wounds with bacterial infections can evolve into bacteremia, sepsis, or septicemia and severe deterioration in wound status. Severely chronic wounds can result in amputation, a frequent occurrence with diabetic foot ulcers, and a major issue in modern healthcare [76,77]. Diabetic patient wound care management must be undertaken with care and rigor. In 2014, chronic wounds impacted 15% of all Medicare beneficiaries in the United States, with an estimated cost of $28–$32 billion annually [78]. Among the 15% of Medicare beneficiaries (8.2 million in population), who had at least one type of wound or infection, surgical infections were most prevalent (4.0%), followed by diabetic infections (3.4%) [79]. The huge cost of nonhealing chronic wounds has become an urgent issue that should be taken seriously.

3.1. The Role of pH Value in Wound Healing

pH value is one of the critical factors involved in the wound healing process for both acute wounds and chronic wounds; it affects matrix metalloproteinase activity, fibroblast activity, keratinocyte proliferation, microbial proliferation, biofilm formation, and immunological responses [80].

Different stages of the wound healing process may require environments of different pH to recover from skin damage and infection. Imbalances in pH can result in serious chronic wounds.

Matrix metalloproteinases (MMPs) are a family of more than 20 proteases. They are able to degrade extracellular components and facilitate the removal of damaged tissues after new tissues have formed. However, excessive protease amounts can interrupt the wound healing process, since they cause endothelial cell damage and degrade the epidermal–dermal junction, which eventually destroys the newly forming tissues [9,80]. To limit expression of MMPs, metalloproteinases (TIMPs) act as inhibitors; they are often at low levels in chronic wounds. Balancing the levels of MMPs and TIMPs is necessary for successful wound healing. Figure 4 (adapted from [9]) depicts the pH-dependent activity of four proteases important in wound healing. Because chronic wounds remain in an alkaline milieu, this could be one cause for high expression of MMPs that lead to unrecovered chronic wounds. However, most proteases can be inactivated in an acid milieu, so decreasing the pH value in wound sites may be an effective method for treating chronic wounds.

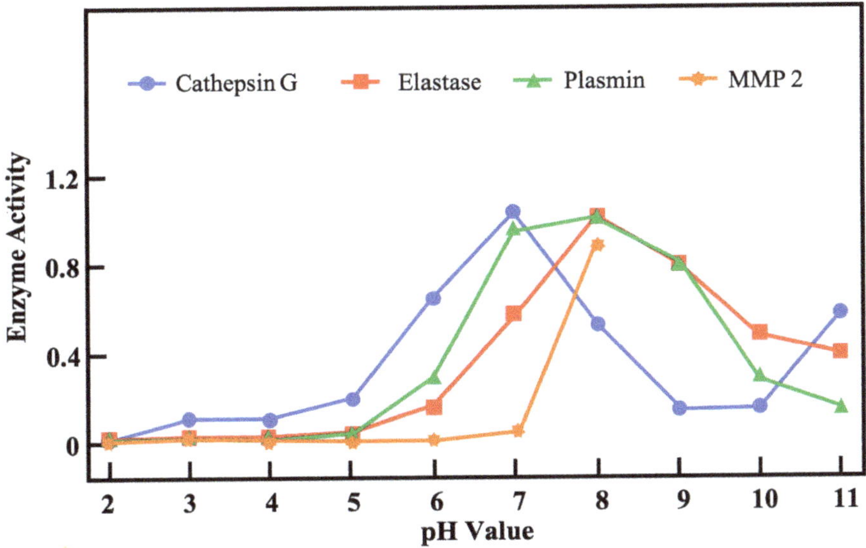

Figure 4. Assessment of pH-dependent activity profiles of four proteases important in wound healing (adapted from [9]).

Lowering the pH of the wound environment may enhance wound contraction. Pipelzadeh et al. [8] stimulated the responses of strips of rat superficial fascia in vitro using physiological solutions at different pH values (5.5, 6.6, 7.3, 8.1) and containing adenosine, calcium and potassium ions, and mepyramine. Their results suggest that there was an at least sixfold increase in the responsiveness to adenosine under acidic conditions (pH 5.5) compared with controls, while in alkaline conditions, responses were not significantly different from control responses.

Lönnqvist et al. [81] found that human keratinocytes cultured in medium with a pH value of 6.0 showed a decreased ability to re-populate the scratched area compared to controls, and no wounds presented re-epithelialization when cultured at pH 5.0. These results suggest that any efforts to alter local wound pH should remain over pH 5.0, which could ensure better re-epithelialization.

It is possible that different stages of wound healing require environments of different pH. An acid milieu could improve fibroblast proliferation [82], whereas neutral and alkaline environments with a pH range of 7.21–8.34 are better for re-epithelialization [83,84]. To prevent the development of chronic wounds, an acid milieu is useful for depressing excessive amounts of MMPs and helps to form the protective acid mantle and normal skin barrier.

3.2. pH in Infected Wounds

Breidenbach et al. [85] showed that tissue microbial levels equal to or higher than 10^4 cfu/g were responsible for delayed wound healing. However, bacterial biofilm can affect the condition of chronic wounds, alter host immune response, affect interactions with other microorganisms, and alter pH, temperature, and nutrient levels in wound regions [86].

Ammonia, a by-product of bacterial metabolism, raises tissue pH [87]. Human pathogenic bacteria grow well when the pH value is above 6, which could lead to more complicated situation for chronic wounds. Because the pH fluctuation in chronic wounds (Figure 3b) is significantly distinct from that of the normal acid mantle, the skin fails to prevent bacterial colonization. Furthermore, the activity of MMPs is heightened when the milieu changes from neutral to alkaline, and infected chronic wounds are subjected to repeated inflammatory states that prevent healing. Changes in pH change can also influence the toxicity of bacterial end products and thus affect their enzyme activity. For instance, lowering pH has been shown to induce structural changes in staphylococcal enterotoxin (SEC2), which has been discussed above [51]. Decreasing pH in wound regions may not only suppress excess MMP expression but also inhibit bacterial growth and toxicity, making it a synergistical method for treating chronic wounds.

Percival et al. [88] reviewed the effect of some antiseptics on pH. In other research, Percival and coworkers [89] found that the antimicrobial performance of silver dressing was significantly enhanced at a pH of 5.5 when compared with a pH of 7.0. It was presumed that the oxidized form of silver present in an acidic environment was active for inhibiting bacterial growth. Molecular iodine (I_2) with its active forms (I_2, H_2OI^+) in solution is calculated to be in the pH range of 3–9 and is thought to have great antimicrobial activity. The greatest antimicrobial effect is found within an acid milieu having a pH between 3 and 6, since H_2OI^+ is the critical biocidal agent most released in this environment [90]. Polyhexamethylene biguanide (PHMB) is a cationic antimicrobial that can inactivate the efflux pump of bacterial cell membranes. The maximal antibacterial effect of PHMB has been shown to occur at a pH range of 5–6 [91]. Chlorhexidine (CHX) is a biguanide cationic detergent similar to PHMB and is more stable at a pH range from 5 to 8 in solution, but its antimicrobial efficacy is optimal in a range of pH 5.5–7.0, unlike PHMB, which is more effective in an acidic milieu [92]. There are a number of other examples not cited in this review, but overall, optimizing pH values can benefit wound healing and inhibit microbial growth.

3.3. Wound Dressings

Wound dressings are barriers that prevent wounds from further mechanical damage or infection. An ideal wound dressing should not only serve as protection but should also be able to absorb wound exudates, maintain a wet environment around the wound, and allow normal transportation of nutrients and gases. There are various types of wound dressings already available in clinical use and on the market, such as plasters, strips, tapes, foams, beads, occlusive films, hydrocolloids, hydrogels, alginate, charcoal dressings, silicone gel, etc.

In recent years, some studies have focused on developing smart wound dressings or bandages different from conventional wound dressings and featuring additional functions, such as radiation-activated drug control-released system, the detection of wound status, and wireless connections between smart wound dressings and mobile phones. Smart wound dressings equipped with pH and temperature sensors can utilize the pH change in wounds to activate drug release for monitoring and treating proposes. Mirani et al. [11] designed a multifunctional hydrogel-based wound dressing capable of colorimetric measurement of pH, indication of bacterial infection, and the release of antibiotic agents at the wound site. They also developed an image analysis application that can record and analyze digital images captured by a smartphone. This facilitates monitoring of treatment strategies that patients can manage at home. In 2018, Mostafalu and coworkers [13] reported a smart bandage for treating and monitoring chronic wounds. This hydrogel-based dressing is loaded with a temperature sensor, a pH sensor, a flexible heater, and a thermo-responsive drug release system. It may be well suited for the treatment of chronic wounds. Moreover, the sensor can be connected to a mobile phone via Bluetooth for visual recording of wound status that allows patients to self-manage wounds. Liu and coworkers [14] developed a smart hydrogel wound patch incorporating modified pH indicator dyes. It showed optimal mechanical properties under different calcium and water concentrations. When pH increased, the color of the hydrogel patches underwent a color transition from yellow to orange and red, which matched the clinically meaningful pH range of chronic or infected wounds. Kiaee and coworkers [15] reported on an electronic wound dressing with active topical drug delivery in response to electrically induced pH change. In basic environments, this dressing released drugs in response to the dehydration process. With this smart dressing, chronic wounds can be systematically and effectively treated.

3.4. Wound Care

Debridement, transplant, negative-pressure wound therapy (NPWT or VAC), and antibiotic ointments are commonly used for wound management. Debridement literally means the removal of dead or infected tissues to improve wound healing. There are several techniques to remove damaged tissues including surgical, mechanical, chemical, autolytic, and maggot-based debridement. Skin transplantation is successfully used to close open chronic wounds by placing autologous grafts or allogeneic grafts on the wound surface. This treatment option can support the rate of epithelialization of the wound areas, and rapid covering of central wound areas with keratinocytes [93]. To investigate the effect of NPWT, Armstrong and coworkers [94] analyzed 162 patients (77 assigned to NPWT and 85 control) in a randomized clinical trial and found that more patients healed in the NPWT group. The speed of wound closure and tissue formation was faster in the NPWT group than in control groups. NPWT seems to be safe and effective for treating complex diabetic foot wounds and could potentially reduce re-amputation cases compared to standard care. Antibiotic ointments are widely used for preventing wound infections. It is a fairly common wound management practice in clinic and personal wound care. However, multidrug-resistant (MDR) pathogens have rapidly emerged due to the abuse of antibiotics and have created a growing a threat to human health.

Bowler and coworkers [95] reviewed infections in various wounds and suggested some associated approaches for wound management. Wound cleansing and surgical debridement are complementary with antimicrobial therapy because they could help reduce the opportunity for infection by reducing microbial load and provide better antibiotic penetration. Moreover, debridement can also avoid blood clots accumulating in tissue debris and consequently reduce the potential for microbial growth. To address the issue of MDR pathogens, nanoparticles have been applied to wound dressings to prevent infection. There are already several review articles discussing the effects of silver and silver nanoparticles in wound management [96–99]. However, because silver is toxic towards mammalian cells, therapeutic window questions remain unanswered. Svitlana Chernousova and Matthias Epple [100] reviewed the toxicity of silver (silver ions and silver nanoparticles) in bacterial, cellular, and animal tests, and examined its impact on the environment. Silver toxicity is associated with particle size, size distribution, morphology, crystallinity, surface functionalization, charge, dose, and concentration. For these reasons, silver should be critically assessed and well-characterized before being used in consumer products, cosmetics, and medical products. Additionally, there is an emerging concern regarding silver-resistant bacteria [101].

Infants have fragile skin and poor skin barriers that result in easy damage. For these reasons, wound care for infants should be particularly careful and rigorous. Strodtbeck and coworkers [102] suggest several strategies for infant wound care that are shown in Figure 5 (adapted from [102]). The scientific rationales for each care strategy are further explained in Strodtbeck's article. Stahl and coworkers [103] reported that cleansing and topical antibacterial use has helped prevent infant open wounds from getting infected from synthetic heterografts, septal patches, valves, conduits, or homograft outflow tracts used during the reconstruction of thoracic wounds. Those prostheses could increase wound complications and the risk of endovascular infection. Bookout and coworkers [104] reported on successful treatment of an infant, a toddler, and an adolescent NPWT. Notably, the infant's labial/buttock wound was completely closed without skin transplantation. They also suggested that adjustments of NPWT should be relative to age, size, and tolerance to therapy.

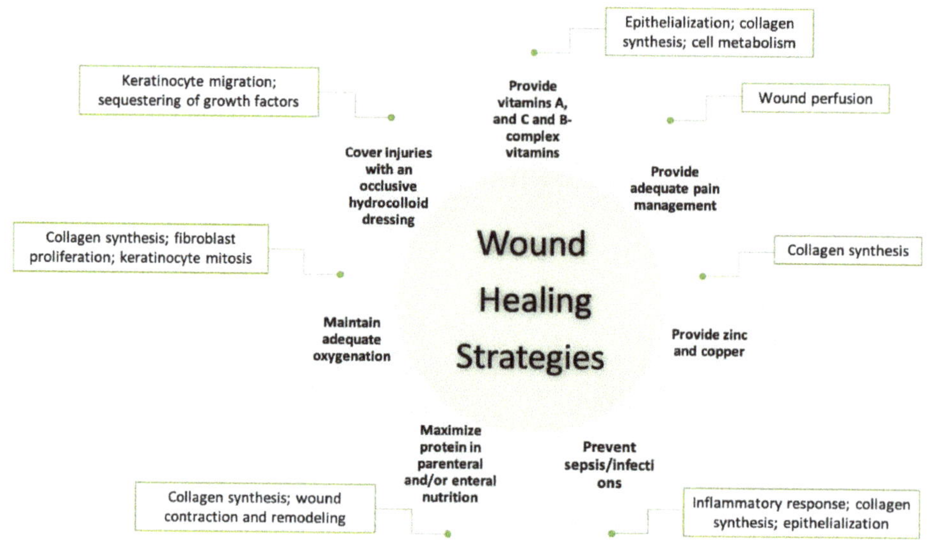

Figure 5. Strategies to promote wound healing in newborns/infants (adapted from [102]).

We suggest monitoring pH change during wound management as a means of evaluating wound healing. Measuring pH value in wound sites is relatively simple and fast. It could be used to provide a rough reference for physicians to diagnose and assess wound healing state, including whether wounds are receiving effective treatment or are becoming chronic, unhealing wounds. Additional examination and analysis, including ultrasound, wound depth measurement, and infection testing could supplement this initial, rough assessment to best determine treatment protocol.

4. Future

pH value is one of the most important factors affecting physiological performance. It may be very useful for making diagnoses and for monitoring and treating wounds and skin-related diseases. pH influences skin barrier functions and plays a critical role in regulating the activation of proteases related to wound healing processes and the development of chronic wounds. Furthermore, it has been correlated with bacterial enterotoxin activity and antibacterial efficacy. Monitoring pH changes in wound regions is a fast and simple method for monitoring wound healing processes and preventing the formation of chronic wounds. Developing a smart bandage equipped with a pH sensor or other systems (e.g., drug control-released carriers) have become increasingly popular in recent years. Besides, tools for non-invasive diagnosis and therapy monitoring in dermatology (e.g., raster-scan optoacoustic mesoscopy (RSOM) and in vivo reflectance confocal microscopy (IVCM)) have also been introduced [105,106]. They are expected to promote additional tools and processes to promote wound healing. While pH value is a simple physiological concept, it could promote a significantly positive impact on not only clinically relevant diagnoses but personal health management in the home.

Author Contributions: Conceptualization, S.-H.K., C.-M.C.; formal analysis, S.-H.K.; writing—original draft preparation, S.-H.K.; writing—review and editing, S.-H.K., C.-M.C., C.-J.S., C.-F.S.; visualization, S.-H.K.; supervision, C.-M.C., C.-J.S., C.-F.S. All authors have read and agreed to the published version of the manuscript.

Funding: The research was funded by the Ministry of Science and Technology, Taiwan (MOST 108-2745-8-007-001 & 109-2623-E-007-002-D).

Acknowledgments: We thank our colleagues who provided insight and expertise that greatly assisted this review.

Conflicts of Interest: The funders had no role in the design of the study; in the collection, analyses, or interpretation of data; in the writing of the manuscript, or in the decision to publish the results.

References

1. Hwa, C.; Bauer, E.A.; Cohen, D.E. Skin Biology. *Dermatol. Ther.* **2011**, *24*, 464–470. [CrossRef] [PubMed]
2. Amsden, B.G.; Goosen, M.F.A. Transdermal Delivery of Peptide and Protein Drugs: An Overview. *AIChE J.* **1995**, *41*, 1972–1997. [CrossRef]
3. Schade, H.; Marchionini, A. Der Säuremantel Der Haut (Nach Gaskettenmessungen). *Klin. Wochenschr.* **1928**, *7*, 12–14. [CrossRef]
4. Sochorová, M.; Staňková, K.; Pullmannová, P.; Kováčik, A.; Zbytovská, J.; Vávrová, K. Permeability Barrier and Microstructure of Skin Lipid Membrane Models of Impaired Glucosylceramide Processing. *Sci. Rep.* **2017**, *7*, 1–8. [CrossRef] [PubMed]
5. Elias, P.M. Epidermal Lipids, Barrier Function, and Desquamation. *J. Invest. Dermatol.* **1983**, *80*, 44–49. [CrossRef] [PubMed]
6. Sharpe, J.R.; Harris, K.L.; Jubin, K.; Bainbridge, N.J.; Jordan, N.R. The Effect of PH in Modulating Skin Cell Behaviour. *Br. J. Dermatol.* **2009**, *161*, 671–673. [CrossRef] [PubMed]
7. Gethin, G. The Significance of Surface PH in Chronic Wounds. *Wounds UK* **2007**, *3*, 52–56.
8. Pipelzadeh, M.H.; Naylor, I.L. The in Vitro Enhancement of Rat Myofibroblast Contractility by Alterations to the PH of the Physiological Solution. *Eur. J. Pharmacol.* **1998**, *357*, 257–259. [CrossRef]
9. Greener, B.; Hughes, A.A.; Bannister, N.P.; Douglass, J. Proteases and PH in Chronic Wounds Proteolytic. *J. Wound Care* **2005**, *14*, 59–61. [CrossRef]
10. Schneider, L.A.; Korber, A.; Grabbe, S.; Dissemond, J. Influence of PH on Wound-Healing: A New Perspective for Wound-Therapy? *Arch. Dermatol. Res.* **2007**, *298*, 413–420. [CrossRef]
11. Mirani, B.; Pagan, E.; Currie, B.; Siddiqui, M.A.; Hosseinzadeh, R.; Mostafalu, P.; Zhang, Y.S.; Ghahary, A.; Akbari, M. An Advanced Multifunctional Hydrogel-Based Dressing for Wound Monitoring and Drug Delivery. *Adv. Healthc. Mater.* **2017**, *6*, 1–15. [CrossRef] [PubMed]
12. Ziaie, B.; Ochoa, M.; Zhou, J.; Jiang, H.; Yoon, C.K.; Oscai, M.; Jain, V.; Morken, T.; Oliveira, R.H.; Maddiplata, D.; et al. A Manufacturable Smart Dressing with Oxygen Delivery and Sensing Capability for Chronic Wound Management. In *Proc. SPIE 10639, Micro- and Nanotechnology Sensors, Systems, and Applications X*; SPIE: Orlando, FL, USA, 2018; pp. 106391C1–106391C10.
13. Mostafalu, P.; Tamayol, A.; Rahimi, R.; Ochoa, M.; Khalilpour, A.; Kiaee, G.; Yazdi, I.K.; Bagherifard, S.; Dokmeci, M.R.; Ziaie, B.; et al. Smart Bandage for Monitoring and Treatment of Chronic Wounds. *Small* **2018**, *14*, 1703509. [CrossRef] [PubMed]
14. Liu, L.; Li, X.; Nagao, M.; Elias, A.L.; Narain, R.; Chung, H.J. A PH-Indicating Colorimetric Tough Hydrogel Patch towards Applications in a Substrate for Smart Wound Dressings. *Polymers (Basel)* **2017**, *9*, 558. [CrossRef] [PubMed]
15. Kiaee, G.; Mostafalu, P.; Samandari, M.; Sonkusale, S. A PH-Mediated Electronic Wound Dressing for Controlled Drug Delivery. *Adv. Healthc. Mater.* **2018**, *7*, 1–8. [CrossRef] [PubMed]
16. Jain, S.; Lahoti, A.; Pandey, K.; Rao, P.K. Evaluation of Skin and Subcutaneous Tissue Thickness at Insulin Injection Sites in Indian, Insulin Naïve, Type-2 Diabetic Adult Population. *Indian J. Endocrinol. Metab.* **2013**, *17*, 864. [CrossRef] [PubMed]
17. Behrendt, H.; Green, M. Skin PH Pattern in the Newborn Infant. *Am. J. Dis. Child.* **1958**, *95*, 35–41. [CrossRef]
18. Garcia Bartels, N.; Scheufele, R.; Prosch, F.; Schink, T.; Proquitté, H.; Wauer, R.R.; Blume-Peytavi, U. Effect of Standardized Skin Care Regimens on Neonatal Skin Barrier Function in Different Body Areas. *Pediatr. Dermatol.* **2010**, *27*, 1–8. [CrossRef]

19. Giusti, F.; Martella, A.; Bertoni, L.; Seidenari, S. Skin Barrier, Hydration, and PH of the Skin of Infants under 2 Years of Age. *Pediatr. Dermatol.* **2001**, *18*, 93–96. [CrossRef]
20. Cleminson, J.; McGuire, W. Topical Emollient for Preventing Infection in Preterm Infants. *Cochrane Database Syst. Rev.* **2016**, *2016*. [CrossRef]
21. Hans Behrendt, M.D.; Marvin Green, M.P. Patterns of Skin PH from Birth through Adolescence, with a Synopsis on Skin Growth. *Amer J. Dis. Child.* **1971**, *122*, 184.
22. Blank, I.H. Patterns of Skin PH from Birth through Adolescence: With a Synopsis on Skin. *Arch Dermatol.* **1971**, *104*, 224–225. [CrossRef]
23. Zlotogorski, A. Distribution of Skin Surface PH on the Forehead and Cheek of Adults. *Arch. Dermatol. Res.* **1987**, *279*, 398–401. [CrossRef] [PubMed]
24. Blank, I.H. Measurement of PH of the Skin Surface. *J. Invest. Dermatol.* **1938**, *2*, 75–79. [CrossRef]
25. Ehlers, C.; Ivens, U.I.; Senderovitz, T.; Serup, J. Females Have Lower Skin Surface PH than Men A. *Ski. Res. Technol.* **2001**, *9*, 90–94. [CrossRef] [PubMed]
26. Luebberding, S.; Krueger, N.; Kerscher, M. Skin Physiology in Men and Women: In Vivo Evaluation of 300 People Including TEWL, SC Hydration, Sebum Content and Skin Surface PH. *Int. J. Cosmet. Sci.* **2013**, *35*, 477–483. [CrossRef]
27. Schmid-Wendtner, M.H.; Korting, H.C. The PH of the Skin Surface and Its Impact on the Barrier Function. *Skin Pharmacol. Physiol.* **2006**, *19*, 296–302. [CrossRef]
28. Issachar, N.; Gall, Y.; Borell, M.T.; Poelman, M.C. PH Measurements during Lactic Acid Stinging Test in Normal and Sensitive Skin. *Contact Dermat.* **1997**, *36*, 152–155. [CrossRef]
29. Itoh, S.; Nakayama, T. Ammonia in Human Sweat and Its Origin. *Jpn. J. Physiol.* **1952**, *3*, 133–137. [CrossRef]
30. Matousek, J.L.; Campbell, K.L. A Comparative Review of Cutaneous PH. *Vet. Dermatol.* **2002**, *13*, 293–300. [CrossRef]
31. Hans, C.K.; Otto, B.-F. The Effect of Detergents on Skin PH and Its Consequence. *Clin. Dermatol.* **1996**, *14*, 23–27.
32. Barel, A.O.; Lambrecht, R.; Clarys, P.; Morrison, B.M.; Paye, M. A Comparative Study of the Effects on the Skin of a Classical Bar Soap and a Syndet Cleansing Bar in Normal Use Conditions. *Skin (Los. Angeles)* **2001**, *7*, 98–104. [CrossRef] [PubMed]
33. Gfatter, R.; Hackl, P.; Braun, F. Effects of Soap and Detergents on Skin Surface PH, Stratum Corneum Hydration and Fat Content in Infants. *Dermatology* **1997**, *195*, 258–262. [CrossRef] [PubMed]
34. Schmid, M.H.; Korting, H.C. The Concept of the Acid Mantle of the Skin: Its Relevance for the Choice of Skin Cleansers. *Dermatology* **1995**, *191*, 276–280. [CrossRef] [PubMed]
35. Grunewald, A.M.; Gloor, M.; Gehring, W.; Kleesz, P. Damage to the Skin by Repetitive Washing. *Contact Dermatitis* **1995**, *32*, 225–232. [CrossRef]
36. Lambers, H.; Piessens, S.; Bloem, A.; Pronk, H.; Finkel, P. Natural Skin Surface PH Is on Average below 5, Which Is Beneficial for Its Resident Flora. *Int. J. Cosmet. Sci.* **2006**, *28*, 359–370. [CrossRef]
37. Grunewald, A.M.; Gloor, M.; Kleesz, P. Barrier Recompensation Mechanisms. *Curr. Probl. Dermatol.* **1996**, *25*, 206–213.
38. Hartmann, A.A. Effect of Occlusion on Resident Flora, Skin-Moisture and Skin-PH. *Dermatol. Res.* **1983**, *275*, 251–254. [CrossRef]
39. Aly, R.; Shirley, C.; Cunico, B.; Maibach, H. Effect of Prolonged Occlusion on the Microbial Flora, PH, Carbon Dioxide and Transepidermal Water Loss on Human Skin. *J. Invest. Dermatol.* **1978**, *31*, 378–381. [CrossRef]
40. Elsner, P.; Maibach, H.I.; Francisco, S. The Effect of Prolonged Drying on Transepidermal Water Loss, Capacitance and PH of Human Vulvar and Forearm Skin. *Acta Derm. Venereol.* **1990**, *70*, 105–109.
41. Waller, J.M.; Maibach, H.I. Age and Skin Structure and Function, a Quantitative Approach (I): Blood Flow, PH, Thickness, and Ultrasound Echogenicity. *Ski. Res. Technol.* **2005**, *11*, 221–235. [CrossRef]
42. Warrier, A.G.; Kligman, A.M.; Harper, R.A.; Bowman, J.; Wickett, R.R. A Comparison of Black and White Skin Using Noninvasive Methods. *J. Cosmet. Sci.* **1996**, *47*, 229–240.
43. Alexis, A.F.; Anigbogu, A.; Baldeweck, T.; Barbosa, V.H.; Bargo, P.R.; Batisse, D.; Berardesca, E.; Brauner, G.J.; de Lacharrière, O.; de Rigal, J.; et al. *Ethnic Skin and Hair*; CRC Press: Boca Raton, FL, USA, 2007; pp. 116–117.
44. Rawlings, A.V. Ethnic Skin Types: Are There Differences in Skin Structure and Function? *Int. J. Cosmet. Sci.* **2006**, *28*, 79–93. [CrossRef] [PubMed]

45. Kim, M.K.; Choi, S.Y.; Byun, H.J.; Huh, C.H.; Park, K.C.; Patel, R.A.; Shinn, A.H.; Youn, S.W. Comparison of Sebum Secretion, Skin Type, PH in Humans with and without Acne. *Arch. Dermatol. Res.* **2006**, *298*, 113–119. [CrossRef] [PubMed]
46. Weiner, J.S.; van Heyningen, R.E. Observations on Lactate Content of Sweat. *J. Appl. Physiol.* **1952**, *4*, 734–744. [CrossRef] [PubMed]
47. Robinson, H.; Robinson, S. Chemical Composition of Sweat. *Physiol. Rev.* **1954**, *34*, 202–220. [CrossRef]
48. Grice, E.A.; Kong, H.H.; Conlan, S.; Deming, C.B.; Davis, J.; Young, A.C.; Bouffard, G.G.; Blakesley, R.W.; Murray, P.R.; Green, E.D.; et al. Topographical and Temporal Diversity of the Human Skin Microbiome. *Science* **2009**, *324*, 1190–1192. [CrossRef]
49. Mark, V.; Dahl, M. Staphylococcus Aureus and Atopic Dermatitis. *Arch Dermatol.* **1983**, *119*, 840–846.
50. Cole, G.W.; Silverberg, N.L. The Adherence of Staphylococcus Aureus to Human Corneocytes. *Arch. Dermatol.* **1986**, *122*, 166–169. [CrossRef]
51. Kumaran, D.; Eswaramoorthy, S.; Furey, W.; Sax, M.; Swaminathan, S. Structure of Staphylococcal Enterotoxin C2 at Various PH Levels. *Acta Crystallogr. Sect. D Biol. Crystallogr.* **2001**, *57*, 1270–1275. [CrossRef]
52. Anderson, D.S. The Acid-Base Balance of the Skin. *Br. J. Dermatol.* **1951**, *63*, 283–295. [CrossRef]
53. Fluhr, J.W.; Elias, P.M. Stratum Corneum PH: Formation and Function of the "Acid Mantle". *Exog. Dermatol.* **2002**, *1*, 163–175. [CrossRef]
54. Beare, J.M.; Cheeseman, E.A.; Gailey, A.A.H.; Neill, D.W. The pH of the Skin Surface of Children with Seborrhoeic Dermatitis Compared with Unaffected Children. *Br. J. Dermatol.* **1958**, *70*, 233–241. [CrossRef] [PubMed]
55. MacKay, B.J.; Denepitiya, L.; Iacono, V.J.; Krost, S.B.; Pollock, J.J. Growth-Inhibitory and Bactericidal Effects of Human Parotid Salivary Histidine-Rich Polypeptides on Streptococcus Mutans. *Infect. Immun.* **1984**, *44*, 695–701. [CrossRef] [PubMed]
56. Lehrer, R.I.; Ladra, K.M.; Hake, R.B. Nonoxidative Fungicidal Mechanisms of Mammalian Granulocytes: Demonstration of Components with Candidacidal Activity in Human, Rabbit, and Guinea Pig Leukocytes. *Infect. Immun.* **1975**, *11*, 1226–1234. [CrossRef]
57. Community, D. Mechanisms for the Microbicidal Activity of Cationic Proteins of Human Granulocytes. *Infect. Immun.* **1976**, *14*, 1269–1275.
58. Chikakane, K.; Takahashi, H. Measurement of Skin PH and Its Significance in Cutaneous Diseases. *Clin. Dermatol.* **1995**, *13*, 299–306. [CrossRef]
59. Berg, R.W.; Milligan, M.C.; Sarbaugh, F.C. Association of Skin Wetness and PH With Diaper Dermatitis. *Pediatr. Dermatol.* **1994**, *11*, 18–20. [CrossRef]
60. Zimmerer, R.E.; Lawson, K.D.; Calvert, C.J. The Effects of Wearing Diapers on Skin. *Pediatr. Dermatol.* **1986**, *3*, 95–101. [CrossRef]
61. Buckingham, K.W.; Berg, R.W. Etiologic Factors in Diaper Dermatitis: The Role of Feces. *Pediatr. Dermatol.* **1986**, *3*, 107–112. [CrossRef]
62. Buffo, J.; Herman, M.A.; Soll, D.R. A Characterization of PH-Regulated Dimorphism in Candida Albicans. *Mycopathologia* **1984**, *85*, 21–30. [CrossRef]
63. Whiteway, M.; Bachewich, C. Morphogenesis in Candida Albicans. *Annu. Rev. Microbiol.* **2007**, *61*, 529–553. [CrossRef] [PubMed]
64. Gow, N.A.R.; Van De Veerdonk, F.L.; Brown, A.J.P.; Netea, M.G. Candida Albicans Morphogenesis and Host Defence: Discriminating Invasion from Colonization. *Nat. Rev. Microbiol.* **2012**, *10*, 112–122. [CrossRef] [PubMed]
65. Whiteway, M.; Oberholzer, U. Candida Morphogenesis and Host-Pathogen Interactions. *Curr. Opin. Microbiol.* **2004**, *7*, 350–357. [CrossRef] [PubMed]
66. Scherwitz, C. Ultrastructure of Human Cutaneous Candidosis. *J. Invest. Dermatol.* **1982**, *78*, 200–205. [CrossRef] [PubMed]
67. Yosipovitch, G.; Tur, E.; Cohen, O.; Rusecki, Y. Skin Surface PH in Intertriginous Areas in NIDDM Patients. *Diabetes Care* **1993**, *16*, 560–563. [CrossRef] [PubMed]
68. Dhar, S. Newborn Skin Care Revisited. *Indian J. Dermatol.* **2007**, *52*, 1–4. [CrossRef]
69. Tyebkhan, G. Skin Cleansing in Neonates and Infants-Basics of Cleansers. *Indian J. Pediatr.* **2002**, *69*, 767–769. [CrossRef]
70. Nguyen, T.H. Textbook of Dermatologic Surgery. *Mayo Clin. Proc.* **1998**, *73*, 709. [CrossRef]

71. SARKAR, R.; BASU, S.; AGRAWAL, R.; GUPTA, P. Skin Care for the Newborn Principles of Skin Care of the Newborn. *Indian Periaterics* **2010**, *47*, 593–598. [CrossRef]
72. Jung, Y.C.; Kim, E.J.; Cho, J.C.; Suh, K.D.; Nam, G.W. Effect of Skin PH for Wrinkle Formation on Asian: Korean, Vietnamese and Singaporean. *J. Eur. Acad. Dermatol. Venereol.* **2013**, *27*, 328–332. [CrossRef]
73. Youn, S.H.; Choi, C.W.; Choi, J.W.; Youn, S.W. The Skin Surface PH and Its Different Influence on the Development of Acne Lesion According to Gender and Age. *Ski. Res. Technol.* **2013**, *19*, 131–136. [CrossRef] [PubMed]
74. Ali, S.M.; Yosipovitch, G. Skin PH: From Basic Science to Basic Skin Care. *Acta Derm. Venereol.* **2013**, *93*, 261–267. [CrossRef] [PubMed]
75. Nix, D.H. Factors to Consider When Selecting Skin Cleansing Products. *J. Wound Cont. Nurses* **2000**, *27*, 260–268.
76. Reiber, G.E.; Vileikyte, L.; Boyko, E.J.; del Aguila, M.; Smith, D.G.; Lavery, L.A.; Boulton, A.J. Causal Pathways for Incident L o w e R- E x t Remity Ulcers in Patients. *Diabetes Care* **1999**, *22*, 157–162. [CrossRef]
77. Han, G.; Ceilley, R. Chronic Wound Healing: A Review of Current Management and Treatments. *Adv. Ther.* **2017**, *34*, 599–610. [CrossRef]
78. Ducker, S. New Study Documents Cost and Impact of Chronic Wounds; Demonstrates Imperative for Wound-Specific Quality Measures, Payment Models and Federal Research Funding. Alliance of Wound Care Stakeholders. 2017. Available online: https://www.woundsource.com/blog/alliance-update-new-study-documents-cost-and-impact-chronic-wounds-demonstrates-imperative (accessed on 14 February 2020).
79. Nussbaum, S.R.; Carter, M.J.; Fife, C.E.; daVanzo, J.; Haught, R.; Nusgart, M.; Cartwright, D. An Economic Evaluation of the Impact, Cost, and Medicare Policy Implications of Chronic Nonhealing Wounds. *Value Heal.* **2018**, *21*, 27–32. [CrossRef]
80. Percival, S.L.; McCarty, S.; Hunt, J.A.; Woods, E.J. The Effects of PH on Wound Healing, Biofilms, and Antimicrobial Efficacy. *Wound Repair Regen.* **2014**, *22*, 174–186. [CrossRef]
81. Lönnqvist, S.; Emanuelsson, P.; Kratz, G. Influence of Acidic PH on Keratinocyte Function and Re-Epithelialisation of Human in Vitro Wounds. *J. Plast. Surg. Hand Surg.* **2015**, *49*, 346–352. [CrossRef]
82. Liu, Y.; Kalén, A.; Risto, O.; Wahlström, O. Fibroblast Proliferation Due to Exposure to a Platelet Concentrate in Vitro Is PH Dependent. *Wound Repair Regen.* **2002**, *10*, 336–340. [CrossRef]
83. Kruse, C.R.; Singh, M.; Targosinski, S.; Sinha, I.; Sørensen, J.A.; Eriksson, E.; Nuutila, K. The Effect of PH on Cell Viability, Cell Migration, Cell Proliferation, Wound Closure, and Wound Reepithelialization: In Vitro and in Vivo Study. *Wound Repair Regen.* **2017**, *25*, 260–269. [CrossRef] [PubMed]
84. Schreml, S.; Szeimies, R.M.; Karrer, S.; Heinlin, J.; Landthaler, M.; Babilas, P. The Impact of the PH Value on Skin Integrity and Cutaneous Wound Healing. *J. Eur. Acad. Dermatol. Venereol.* **2010**, *24*, 373–378. [CrossRef] [PubMed]
85. Breidenbach, W.C.; Trager, S. Quantitative Culture Technique and Infection in Complex Wounds of the Extremities Closed with Free Flaps. *Plast. Reconstr. Surg.* **1995**, *95*, 860–865. [CrossRef] [PubMed]
86. Percival, S.L.; Thomas, J.G.; Williams, D.W. Biofilms and Bacterial Imbalances in Chronic Wounds: Anti-Koch. *Int. Wound J.* **2010**, *7*, 169–175. [CrossRef]
87. Leveen, H.H.; Falk, G.; Borek, B.; Diaz, C.; Lynfield, Y.; Wynkoop, B.J.; Mabunda, G.A.; Rubricius, J.L.; Christoudias, G.C. Chemical Acidification of Wounds. An Adjuvant to Healing and the Unfavorable Action of Alkalinity and Ammonia. *Ann. Surg.* **1973**, *178*, 745–753. [CrossRef]
88. Percival, S.L.; Finnegan, S.; Donelli, G.; Vuotto, C.; Rimmer, S.; Lipsky, B.A. Antiseptics for Treating Infected Wounds: Efficacy on Biofilms and Effect of PH. *Crit. Rev. Microbiol.* **2016**, *42*, 293–309. [CrossRef]
89. Percival, S.L.; Thomas, J.G.; Slone, W.; Linton, S.; Corum, L.; Okel, T. The Efficacy of Silver Dressings and Antibiotics on MRSA and MSSA Isolated from Burn Patients. *Wound Repair Regen.* **2011**, *19*, 767–774. [CrossRef]
90. Wutzler, P.; Sauerbrei, A.; Klöcking, R.; Brögmann, B.; Reimer, K. Virucidal Activity and Cytotoxicity of the Liposomal Formulation of Povidone-Iodine. *Antivir. Res.* **2002**, *54*, 89–97. [CrossRef]
91. Broxton, P.; Woodcock, P.M.; Heatley, F.; Gilbert, P. Interaction of Some Polyhexamethylene Biguanides and Membrane Phospholipids in Escherichia Coli. *J. Appl. Microbiol.* **1984**, *57*, 115–124.
92. Russell, A.D.; Path, F.R.C. Chlorhexidine: Antibacterial Action and Bacterial Resistance. *Infection* **1986**, *14*, 212–215. [CrossRef]

93. Hierner, R.; Degreef, H.; Vranckx, J.J.; Garmyn, M.; Massagé, P.; van Brussel, M. Skin Grafting and Wound Healing—The "Dermato-Plastic Team Approach". *Clin. Dermatol.* **2005**, *23*, 343–352. [CrossRef] [PubMed]
94. Armstrong, D.G.; Lavery, L.A. Negative Pressure Wound Therapy after Partial Diabetic Foot Amputation: A Multicentre, Randomised Controlled Trial. *Lancet* **2005**, *366*, 1704–1710. [CrossRef]
95. Clinmicrorevs, M.; Bowler, P.; Armstrong, D.G. Wound Microbiology and Associated Approaches to Wound. *Clin. Microbiol. Rev.* **2001**, *14*, 244–269.
96. Atiyeh, B.S.; Costagliola, M.; Hayek, S.N.; Dibo, S.A. Effect of Silver on Burn Wound Infection Control and Healing: Review of the Literature. *Burns* **2007**, *33*, 139–148. [CrossRef]
97. Fong, J.; Wood, F. Nanocrystalline Silver Dressings in Wound Management: A Review. *Int. J. Nanomed.* **2006**, *1*, 441–449. [CrossRef]
98. Wilkinson, L.J.; White, R.J.; Chipman, J.K. Silver and Nanoparticles of Silver in Wound Dressings: A Review of Efficacy and Safety. *J. Wound Care* **2011**, *20*, 543–549. [CrossRef]
99. Leaper, D.J. Silver Dressings: Their Role in Wound Management. *Int. Wound J.* **2006**, *3*, 282–294. [CrossRef]
100. Chernousova, S.; Epple, M. Silver as Antibacterial Agent: Ion, Nanoparticle, and Metal. *Angew. Chem. Int. Ed.* **2013**, *52*, 1636–1653. [CrossRef]
101. Percival, S.L.; Bowler, P.G.; Russell, D. Bacterial Resistance to Silver in Wound Care. *J. Hosp. Infect.* **2005**, *60*, 1–7. [CrossRef]
102. Strodtbeck, F. Physiology of Wound Healing. *Nat. Immunol.* **2001**, *1*, 43–52. [CrossRef]
103. Stahl, R.S.; Kopf, G.S. Reconstruction of Infant Thoracic Wounds. *Plast. Reconstr. Surg.* **1988**, *82*, 1000–1011. [CrossRef] [PubMed]
104. Bookout, K.; McCord, S.; McLane, K. Case Studies of an Infant, a Toddler, and an Adolescent Treated with a Negative Pressure Wound Treatment System. *J. Wound Ostomy Cont. Nurs.* **2004**, *31*, 184–192. [CrossRef] [PubMed]
105. Cinotti, E.; Gergelé, L.; Perrot, J.L.; Dominé, A.; Labeille, B.; Borelli, P.; Cambazard, F. Quantification of Capillary Blood Cell Flow Using Reflectance Confocal Microscopy. *Ski. Res. Technol.* **2014**, *20*, 373–378. [CrossRef]
106. Hindelang, B.; Aguirre, J.; Schwarz, M.; Berezhnoi, A.; Eyerich, K.; Ntziachristos, V.; Biedermann, T.; Darsow, U. Non-Invasive Imaging in Dermatology and the Unique Potential of Raster-Scan Optoacoustic Mesoscopy. *J. Eur. Acad. Dermatol. Venereol.* **2019**, *33*, 1051–1061. [CrossRef] [PubMed]

Technical Note

Optical Density Optimization of Malaria Pan Rapid Diagnostic Test Strips for Improved Test Zone Band Intensity

Prince Manta [1], Rupak Nagraik [2], Avinash Sharma [2], Akshay Kumar [3], Pritt Verma [4], Shravan Kumar Paswan [4], Dmitry O. Bokov [5], Juber Dastagir Shaikh [6], Roopvir Kaur [7], Ana Francesca Vommaro Leite [8], Silas Jose Braz Filho [8], Nimisha Shiwalkar [9], Purnadeo Persaud [10] and Deepak N. Kapoor [1,*]

1 School of Pharmaceutical Sciences, Shoolini University of Biotechnology and Management Sciences, Solan 173212, India; princemanta@gmail.com
2 School of Bioengineering and Food Technology, Shoolini University of Biotechnology and Management Sciences, Solan 173212, India; rupak.nagraik@gmail.com (R.N.); avinashsubms@gmail.com (A.S.)
3 Department of Surgery, Medanta Hospital, Gurugram 122001, India; drakshay82@gmail.com
4 Departments of Pharmacology, CSIR-National Botanical Research Institute, Lucknow 226001, India; preetverma06@gmail.com (P.V.); paswanshravan@gmail.com (S.K.P.)
5 Institute of Pharmacy, Sechenov First Moscow State Medical University,8 Trubetskaya St., Moscow 119991, Russia; bokov_d_o@staff.sechenov.ru
6 Department of Neurology, MGM Newbombay Hospital, Vashi, Navi Mumbai 400703, India; jubershaikh703@yahoo.com
7 Department of Anesthesiology, Government Medical College, Amritsar 143001, India; roopvirsaini@gmail.com
8 Department of Medicine, University of Minas Gerais, Passos 37902-313, Brazil; francescavommaroleite@gmail.com (A.F.V.L.); silasbrazf@gmail.com (S.J.B.F.)
9 Department of Anesthesiology, MGM Hospital, Navi Mumbai 410209, India; dr.nimisha4u@gmail.com
10 Department of Medicine, Kansas City University, Kansas City, MO 64106, USA; narpaulpersaud@hotmail.com
* Correspondence: deepakpharmatech@gmail.com

Received: 30 July 2020; Accepted: 14 September 2020; Published: 29 October 2020

Abstract: For the last few decades, the immunochromatographic assay has been used for the rapid detection of biological markers in infectious diseases in humans and animals The assay, also known as lateral flow assay, is utilized for the detection of antigen or antibody in human infectious diseases. There are a series of steps involved in the development of these immuno-chromatographic test kits, from gold nano colloids preparation to nitrocellulose membrane coating (NCM). These tests are mostly used for qualitative assays by a visual interpretation of results. For the interpretation of the results, the color intensity of the test zone is therefore very significant. Herein, the study was performed on a malaria antigen test kit. Several studies have reported the use of gold nanoparticles (AuNPs) with varying diameters and its binding with various concentrations of protein in order to optimize tests. However, none of these studies have reported how to fix (improve) test zone band intensity (color), if different sized AuNPs were synthesized during a reaction and when conjugated equally with same amount of protein. Herein, different AuNPs with average diameter ranging from 10 nm to 50 nm were prepared and conjugated equally with protein concentration of 150 μg/mL with $K_D = 1.0 \times 10^{-3}$. Afterwards, the developed kits' test zone band intensity for all different sizes AuNPs was fixed to the same band level (high) by utilization of an ultraviolet-visible spectrophotometer. The study found that the same optical density (OD) has the same test zone band intensity irrespective of AuNP size. This study also illustrates the use of absorption maxima (λ max) techniques to characterize AuNPs and to prevent wastage of protein while developing immunochromatographic test kits.

Keywords: lateral flow assay; immuno-chromatographic; gold nanoparticles sensor; UV/Vis spectrophotometer; malaria pan rapid diagnostic strip; point-of-care

1. Introduction

Malaria is caused by parasites that are transmitted to humans via the bites of the infected female Anopheles mosquito. While preventable and curable, it still remains a paramount cause of morbidity and mortality in developing countries. Malaria is estimated to kill between 1.5 to 2.7 million people annually [1]. Malaria morbidity is estimated at about 300–500 million annually, and malaria clinical diagnosis is most effective at 50%. Malaria immunoassays use the inherent sensitivity, specificity and binding affinity of antibodies to respective antigens for the detection of antigens in a sample. In immunoassays, the sample tested includes whole blood, urine, saliva, serum, etc. [2]. In the Malaria Pan Antigen rapid test kit, the sample used is Red Blood cells containing specific antigens of *P. vivex and P. malariae/P. ovale* [3]. The red blood cells get lysed by a buffer solution to allow antigen–antibody binding at the test site. Immunoassay signals emanate from the gold-labeled antibody set for the antigen on a substratum at the binding site (Test line). Typical antibody labels include fluorescent molecules, nano- or microparticles, or enzymes. Gold nanoparticles (NPs) are the most widely used label [4]. Such immunoassays can be used in industry, clinical or laboratory settings, doctor's offices, or as over-the-counter tests [2]. At the test line, the naked eye will see a gold-labelled marker as a pink/red line [5]. In most countries, the diagnosis of malaria challenges multiple laboratories [3]. The laboratories require longer than one hour to analyze the findings, leading to less consistency in the analysis of the results.

1.1. Components of Immuno-Chromatographic Test Kits

The Immuno-Chromatographic kit is composed of components shown in Figure 1. The parts of the kits are attached on an inert polyvinyl chloride (PVC) backing material and further packed in a plastic cassette with a specimen port and reaction window displaying the capture and control zones [2]. The Immunochromatographic Test Kit has a sample pad, conventionally composed of glass fibres. The sample pad is selected to have zero cross-reactivity with the specimen. The sample pad is pretreated with a buffer for specimen pH adjustment and extraction of unspecific antigen form specimens [6]. One of the vital parts of the strip is nitrocellulose membrane (NCM). In this, the interaction between antigen and antibody takes place. Typically, a hydrophobic nitrocellulose membrane is used on which anti-target analyte antibodies are immobilized in a line that crosses the membrane to act as a capture zone on the test line [2]. The NCM membrane should be chosen based upon pore size [7]. Other parts of test strips are glass fibres or non-woven fibres based conjugate pads which can be pre-treated to avoid any cross-reactivity [8]. Conclusively, the conjugate pad is prepared by dipping the glass fibers into a colloidal solution of gold protein and then used after drying. In addition, an absorbent pad is present in the kit, which is designed to collect extra specimen samples passing the reaction membrane [9].

1.2. The Protein

In the Malaria Pan immunoassay, antibody protein is used for AuNP conjugation. Plasmodium lactate dehydrogenase (pLDH) and goat anti-mouse (GAM) protein are used for binding at test and control lines, respectively. An ultraviolet-visible spectrophotometer optimization technique was demonstrated in this work by formulating an immuno-chromatographic detection kit for Malaria Pan using AuNPs as an indicator. Various research works attempted to optimize the AuNP size [10–13], and the AuNPs of about 30–40 nm were reported to be optimal [11,12]. Khlebtsov and Byzova et al. also tried to determine the optimum concentration of protein required for AuNP conjugation [14,15].

In present research, gold nanoparticles (AuNPs) were utilized as labels, and the concentration of AuNPs with conjugate antibodies was tailored to a fine-tuned optical density (OD). The gold nanoparticles of various sizes (10 nm to 50 nm) were prepared, by quantifying λ max (absorption maxima) and dynamic light scattering (DLS). The relationship of AuNP diameters with a concentration of target protein was monitored to develop a better test kit. Finally, the developed immuno-chromatographic test kit test zone band intensity was tested using RGB and HSV color models. The reason to select a malaria test kit for the study is to create a more cost-effective rapid diagnostic test kits because malaria cases are found in countries where cost-effectiveness is significant. The study aim to improve test band intensity irrespective of AuNP size using a fixed quantity of protein while optimizing the optical density.

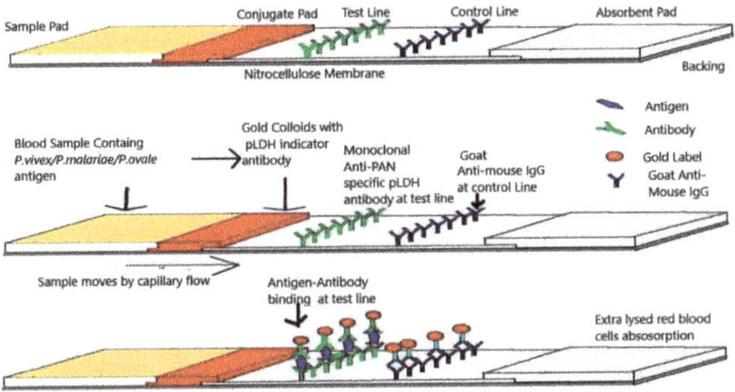

Figure 1. Presentation of lateral flow strip that works on sandwich assay. Blood sample lysed with buffer solution is added to the sample pad. *P. vivex/P. malariae/P. ovale* malaria antigens attach to antibodies in the red colored gold conjugate pad and the complex formed attaches to test line monoclonal anti-PAN specific pLDH antibodies. The excess labeled antibodies bind with Goat anti-mouse IgG antibodies in the control line. The extra lysed red blood cells get absorbed in the absorbent pad.

2. Materials and Methods

2.1. Reagents, Instruments and Other Support Materials

For the fabrication of immunochromatographic strip assay, sodium hydrogen phosphate, sucrose, disodium hydrogen phosphate, sodium chloride and bovine serum albumin (BSA) were purchased from Merck, Darmstadt, Germany. The gold chloride used for the synthesis of gold nanoparticles was purchased from Sigma-Aldrich, Tokyo, Japan. The Plasmodium lactate dehydrogenase (pLDH) antibodies' molecules and control line Goat anti-mouse protein were purchased from Fapon Biotech, Shenzhen, China. All the other chemicals and reagents used in the present study were of analytical grade reagents. The Delsa™ Nano Submicron Particle Size Zeta Potential instrument of Beckman Coulter, Brea, CA, USA was used for analyzing AuNP diameters. An ultraviolet-visible spectrophotometer 1900i of Shimadzu, Kyoto, Japan was used to measure optical density and absorbance maxima. The nitro cellulose membrane was coated with XYXYZ3210™ dispense platform of Bio-dot, Irvine, CA, USA. The centrifuge of Remi RM-12C, Mumbai, India was utilized for AuNP–protein conjugate centrifugation, and the magnetic stirrer of Remi, Mumbai, India was also used in the study. The nitrocellulose membrane was purchased form Nupore System Pvt. Ltd., Ghaziabad, India and Glass fibre sample pad and conjugate pad were purchased from Advanced Micro Devices, Ambala, India.

2.2. Experimental

2.2.1. Preparation of Gold Nanoparticles (AuNPs)

The gold nanoparticles were prepared by classical classical Turkevich and Fern methods by citrate reduction. In general, the Turkevich and Fern process reaction leads to formation of AuNPs of size range 10 nm to 100 nm [16,17]. Utilizing raw gold chloride ($AuHCl_4$) [18], the 1% light yellowish color solution was prepared by dissolving 1 gm of gold chloride in 100 mL of ultra-mili-Q water. The aforementioned 1% gold solution was furthermore dissolved into ultra-milli-Q water in order to obtain the optical density between 0.7 and 0.9 at λ max (absorption maxima) by taking a solution spectrum scan at wavelength between 700 nm and 400 nm. Now, the final gold solution had been refluxed for 30 min at 100 °C. The above-diluted gold solution was reduced by adding 1% (1 g of sodium citrate in 100 mL of water) sodium citrate solution of pH 7.80 ± 0.5 with refluxing until bright red color develops. Initially, the addition of a 1% solution of sodium citrate turns the color of the solution black. The color change from mildly yellowish to brick red confirms the synthesis of nanoparticles. The change in colour solution is due to the surface plasmon resonance effect (SPR) in which electrons excited to its higher state and produces a colour change. During the reduction mechanism, metal salts get converted into their ionic form when it combines with water. Different chemical functional groups of reducing agents combine with metal ions whether they are bivalent or monovalent and reduce it into a zerovalent state of small size [19].

The color transforms from red to pink to blue as the reaction proceeds. As the solution color turns pink, the reaction was stopped by decreasing the temperature to room temperature in the ice bath. Particle size distribution of synthesized nanoparticles was analysed by dynamic light scattering.

The five gold nanoparticles of the sizes 10 nm, 20 nm, 30 nm, 40 nm and 50 nm were chosen for protein conjugation after AuNP characterisation. The prepared pink-colored gold solution was characterized by spectrophotometric absorbance maxima (λ max) by scanning in a visible wavelength range of 700 nm to 400 nm.

2.2.2. Protein Conjugation with Gold NPs

For all the above-prepared AuNPs of size 10 nm to 50 nm, pH was adjusted separately to 7.00 ± 0.1 with 0.2 M Potassium Carbonate solution of pH 12.00 ± 0.5. The pH adjustment is predicated on the protein's isoelectric point, which varies from protein to protein. The pH was measured with the help of pH paper. The antibody pLDH (Plasmodium lactate dehydrogenase) reagents have been diluted to 150 µg/mL from the stock solution with 10 mM of Sodium Dihydrogen Phosphate buffer of pH 8.50 ± 0.1. Afterwards, the protein pLDH of 150 µg/mL concentration was conjugated to all five AuNPs (10 nm to 50 nm). Conjugation of the AuNPs and protein was achieved by stirring the solution to 10 ± 2 min. Following this, 1% BSA (Bovine Serum Albumin) was added into the gold conjugate solution and stirred for 30 ± 2 min for stabilisation and abstraction of unbound protein. For all five sizes of AuNPs, the single tuned protein concentration was used to detect the effect of gold nanoparticle size on the band intensity of developed kits.

2.2.3. Centrifugation

The above five separately prepared AuNP–protein conjugate solutions were centrifuged. The centrifugation was performed with a Relative Centrifugal Force (RCF) of 7000× *g* for 45 min at 4 °C to 8 °C temperature. The centrifugation of an AuNPs–protein conjugate solution at a force higher than 7000× *g* RCF can sometimes shows aggregation while centrifugation at a force slower than 7000× *g* RCF may give less residue with a dark supernatant. Centrifugation at 7000× *g* RCF gives a stable and good yield of the residue or pellet. The supernatant's aspiration was performed in a different beaker, and gold pellets were resuspended in the phosphate-buffered saline (PBS) buffer. The absorbance of the supernatant was measured at a 520 nm wavelength. If the OD was greater than 0.05, then the supernatant's re-centrifugation is performed one more time. The supernatant was discarded if the OD was less than 0.05. This will enable conjugate recovery and prevent wastage. The supernatant aspiration is performed in a separate beaker, accumulating the AnNP–protein pellets. Carefully, the supernatant aspiration and the residue re-suspension was accomplished in a re-suspension buffer. Figure 2 represents the protein conjugation and centrifugation procedure.

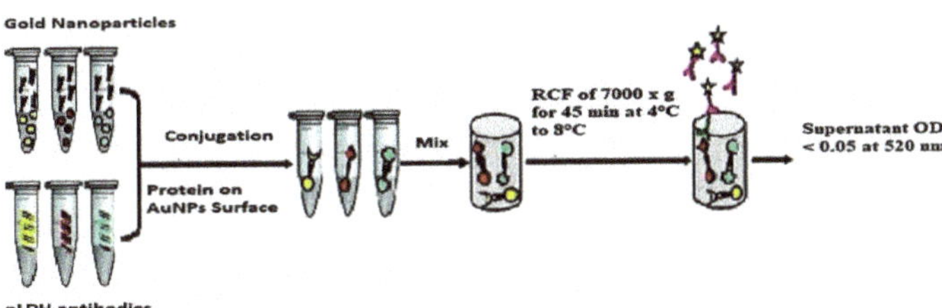

Figure 2. The diagrammatical representation of the protein conjugation and centrifugation methodology.

2.2.4. Conjugate Pad Prepration

All five separately re-suspended AuNPs—protein conjugate above solutions were diluted (ultra-pure mili-Q water) to a constant OD of 3.00 at 520 nm wavelength.

Following the dilution, five conjugate pads were prepared by dipping the glass fibre pad into the conjugate solution. In comparison, the other changes through the entire development of the kit was held constant, e.g., test line concentration and control line protein.

2.2.5. Membrane Coating

The five nitrocellulose membranes were coated at the test (Pan) and control line (C). The test and control line coating on the nitrocellulose membrane (NCM) was achieved with the use of a Bio-dot dispensing machine. First of all, the bio-dot machine stripping system (tubing and jets) was flushed with de-ionized water over ten cycles. The control and test solutions were then coated on NCMs. For drying, membrane sheets were kept in the oven at 30 °C for 30 min after coating. The concentrations of test and control line reagents were as follows:

2.2.6. Test Line Reagents Concentration

To obtain the final test solution, the pLDH (Plasmodium lactate dehydrogenase) antibody was diluted from the stock solution to 50 µg/mL with 1% sucrose solution in the PBS buffer. The antibody protein mixing in PBS buffer was performed with a magnetic stirrer, and a 0.22-micron filter was used to eliminate the suspended particles.

2.2.7. Control Line Reagents Concentration

To obtain the final control line solution, Goat Anti Mouse IgG was diluted from stock solution to 400 µg/mL with a 0.5 percent sucrose in the PBS buffer. To extract the suspended particles, mixing and filtration were achieved using a 0.22-micron filter.

3. Results and Discussion

3.1. Gold NP Characterization

Firstly, we prepared the most stable AuNPs of an average of of 10 nm to 50 nm in diameter. The AuNPs with absorbance maxima (λ max) ranging between 520 to 570 wavelengths were considered for the development of the Malaria Pan Antigen detection test kit. In this range of λ max, the AuNP size ranges from 10 nm to 50 nm as determined by the particle sizer. Figure 3A–E represent the AuNP size measured in dynamic light scattering (DLS). AuNPs of this range were selected due to their smaller particle size and lower polydispersity index (PDI). It has been observed that smaller sized nanoparticles have better conjugation with the protein [10–12,20]. The size of gold NPs depends on the sodium citrate content used [18]. The concentration of sodium citrate in gold solution affects the size of AuNPs, which can be controlled by measuring absorbance maxima (λ max). The sodium citrate of viz 0.2, 0.4, 0.6, 0.75 and 0.90 mg/mL in gold solutions produces nanoparticles (NPs) with average diameters of 10 nm (size distribution of 8 to 12 nm), 20 nm (size distribution of 17 to 23 nm),

30 nm (size distribution of 26 to 35 nm), 40 nm (size distribution of 32 to 50 nm) and 50 nm (size distribution of 36 to 80 nm), respectively (Figure 3). AuNPs shows λ max at varying wavelengths of 520 (10 nm), 530 (20 nm), 540 (30 nm), 560 (40 nm) and 570 (60 nm). Figure 4A–E represent the λ max of AuNPs. The optical density of prepared AuNPs at λ max ranges between 8.0 to 9.0. When sodium citrate concentration increases in AuNP solution, AuNP size does too. When the size of the gold NPs increases, the absorbance maxima (λ max) shift to higher wavelengths (Figure 5), and the color of the solution turns from pink to blue, reflecting nanoparticles' instability.

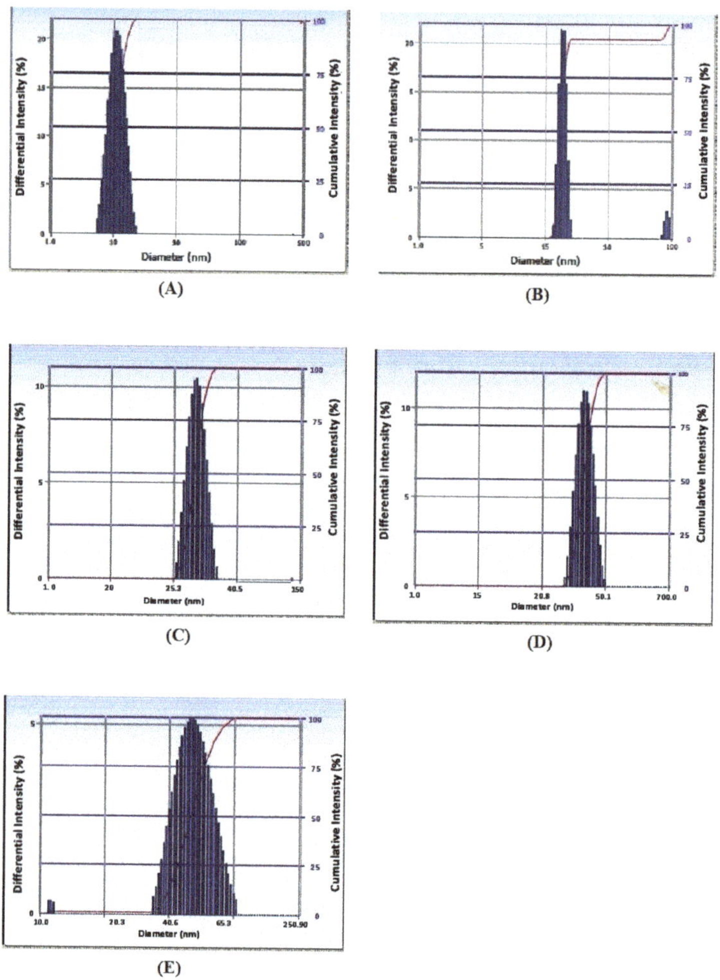

Figure 3. Synthesized gold nanoparticles' size distribution measured by the Zeta Seizer. The sodium citrate solutions of 0.2, 0.4, 0.6, 0.75 and 0.90 mg/mL in gold solutions' produced nanoparticles (NPs) with average diameters of (**A**) 10 nm (size distribution of 8 to 12 nm); (**B**) 20 nm (size distribution of 17 to 23 nm); (**C**) 30 nm (size distribution of 26 to 35 nm); (**D**) 40 nm (size distribution of 32 to 50 nm); and (**E**) 50 nm (size distribution of 36 to 80 nm).

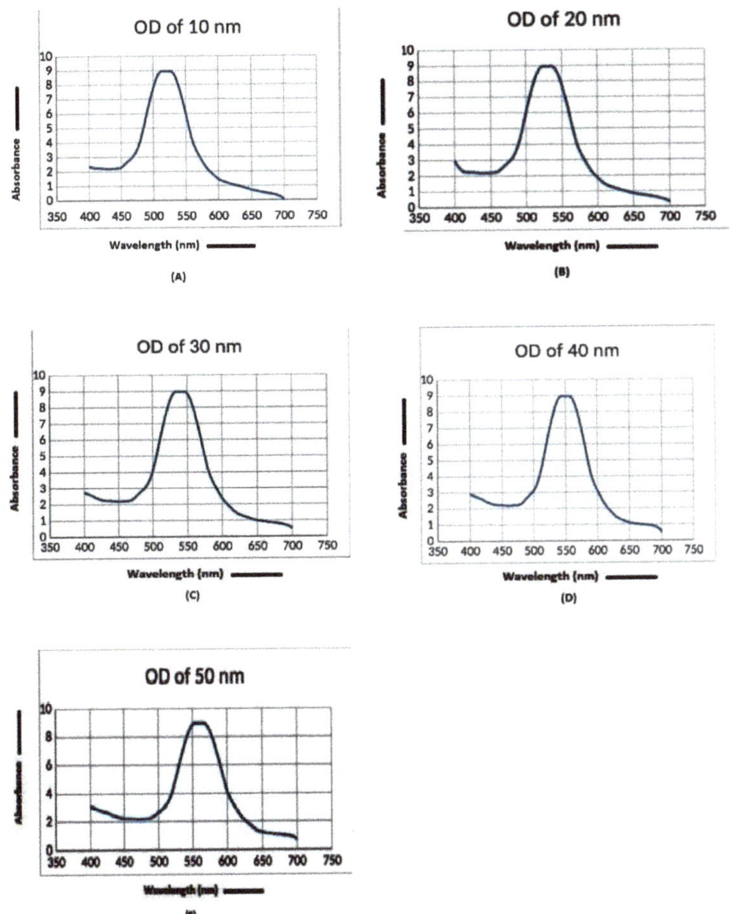

Figure 4. Optical density spectrum of synthesized gold nanoparticles. The gold nanoparticles (AUNPs) show different λ max (absorbance maxima) at different sizes; i.e., **(A)** λ max 520 nm at diameter of 10 nm; **(B)** λ max 530 nm at diameter of 20 nm; **(C)** λ max 540 nm at diameter of 30 nm; **(D)** λ max 560 nm at diameter of 40 nm; **(E)** λ max 570 nm at diameter of 50 nm.

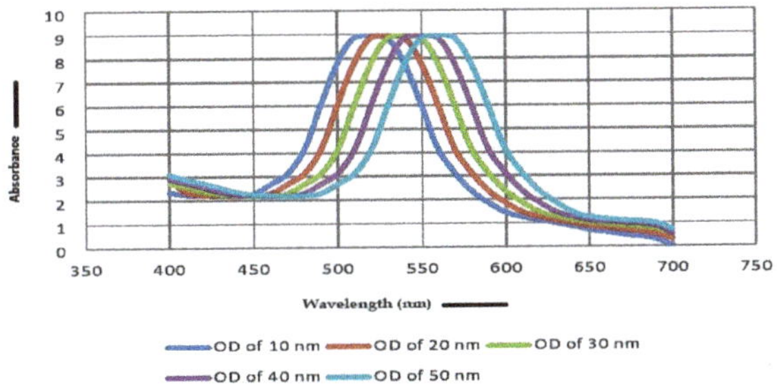

Figure 5. The graph shows a correlation of synthesized gold nanoparticles (10 nm, 20 nm, 30 nm, 40 nm and 50 nm) λ max (absorbance maxima) and Optical Density. With an increase in the size of the gold NPs, the absorbance maxima (λ max) shifts to higher wavelengths (520 nm to 570 nm), reflecting nanoparticle instability.

As the average diameters increase, the nanoparticles' size distribution increases, which causes the instability of AuNPs (Figure 3). This technique is widely used for the determination of particle size in colloidal solution, which, in turn, used to measure the thickness of capping or stabilizing agent along with its actual size of metallic core. These studies also determined the hydrodynamic diameter of the synthesized nanoparticles. These results also suggested that there is an absence of large aggregates when these nanoparticles were dispersed in aqueous medium [21].

3.2. Monitoring the Protein Loss

After centrifigation of AuNP–protein conjugate, a small portion of re-suspended solution was diluted (1 in 100) into ultra-pure mili-Q water to facilitate OD measurement at 520 nm. The OD values obtained are shown in Table 1.

Table 1. Optical density (OD) measurement results of synthesized gold nanoparticles–protein centrifuged re-suspension conjugate solution when diluted in 1–100.

S. No.	Gold Nanoparticles (AuNP) Size	Optical Density (OD) Observation
1	10 nm	0.301
2	20 nm	0.354
3	30 nm	0.368
4	40 nm	0.426
5	50 nm	0.385

Figure 6 and Table 1 indicate that 40 nm AuNPs have high OD, and thus AuNPs have maximum protein binding with AuNPs of 40 nm size. In contrast, the other AuNPs had less binding. This indicates that the supernatant lost extra unbound protein. This means that the additional unbound protein was lost in the supernatant. The protein depletion can also be checked with the use of UV/Vis spectrophotometer OD analysis.

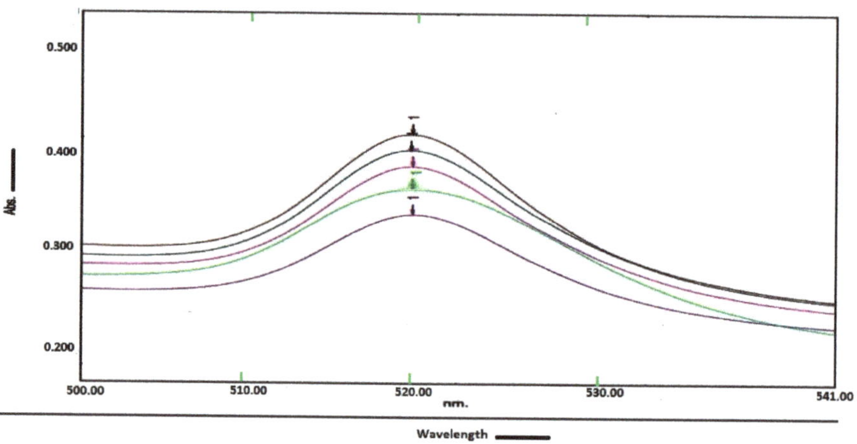

Figure 6. The spectrum overlay graph shows optical density of gold nanoparticles (10 nm, 20 nm, 30 nm, 40 nm and 50 nm) and protein conjugate solution (1–100 mL) measured at 520 nm of wavelength. The gold nanoparticles of average diameter of 40 nm have maximum optical density (0.426), which reflects the maximum protein binding.

3.3. Test Line Intensity Analysis

The developed kits were tested to find out the kit test zone band intensity (results) when equivalent protein ratios are conjugated with different AuNP sizes. The five immunochromatographic rapid test kits were formulated using a conjugate pad (10 nm to 50 nm) prepared above. Then, all five of the immunechromatographic test kits were assembled. Now, the five developed test kits were tested for band intensity using 5 µL of malaria (*P. vivax*) positive blood specimens of concentration 150 parasites/µL. The three specifications were given to test line viz. high test line intensity was ranked as +3, medium test line intensity was ranked as +2, and weak test line intensity was ranked as +1. Upon testing all kits with the same specimen samples, all kits showed an equal band intensity of +3 (high) as shown in Table 2. In this analysis, it was noticed that, if the final OD is tuned to one point, there will be no effects of AuNPs sizes on kit results (test zone intensity). OD-adjustment will refine the final test zone band intensity. Figure 7 is the systematic representation of the results. All five test kits developed after OD tuning to 3.0 were additionally verified for their specificity with 5 µL of Malaria Pf (Falciparum) antigen blood specimens of concentration 40 parasites/µL to find out test kit susceptibility for *P. Falciparum*. The specificity results of developed kits (Table 3) were found without any false positive indication (no Pan line appears), and the the control band intensity was high (+3).

Table 2. Test line intensity results of five Malaria Pan Ag immunochromatographic rapid test kits formulated using different diameter AuNPs. The kit final OD is tuned to one point (3.00) after conjugation with protein. Kit's test line intensity was tested using *P. vivax* positive blood specimen of a concentration of 150 parasites/µL. The band intensity for test (Pan) and control (C) line is ranked as: high test (+3), medium (+2) and weak (+1). Upon testing, all developed kits showed an equal sensitivity of +3 (high).

Test Zone Intensity of Kit Developed by 10 nm of AuNPs	Test Zone Intensity of Kit Developed by 20 nm of AuNPs	Test Zone Intensity of Kit Developed by 30 nm of AuNPs	Test Zone Intensity of Kit Developed by 40 nm of AuNPs	Test Zone Intensity of Kit Developed by 50 nm of AuNPs
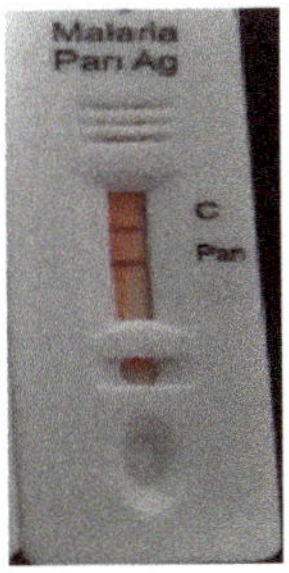 Malaria Pan Ag, C, Pan	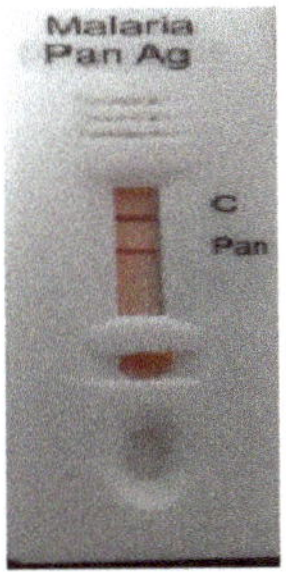 Malaria Pan Ag, C, Pan	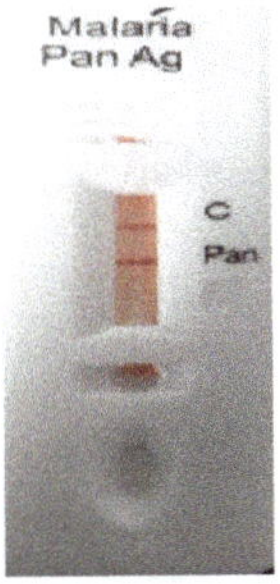 Malaria Pan Ag, C, Pan	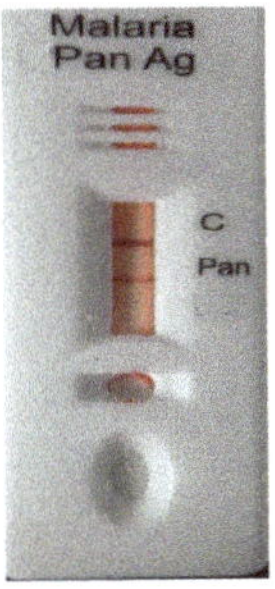 Malaria Pan Ag, C, Pan	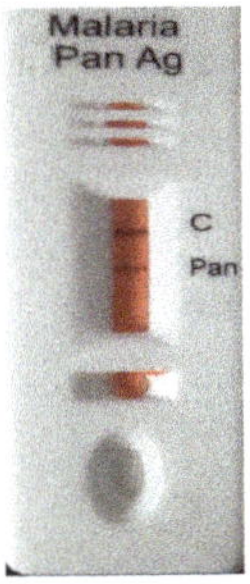 Malaria Pan Ag, C, Pan
+3 (High) Intensity	+3 (High) Intensity	+3 (High) Intensity	+3 (High) Intensity	+3 (High) Intensity

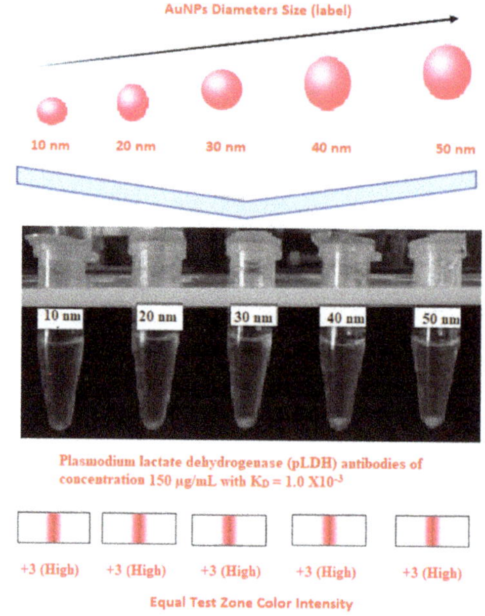

Figure 7. The diagram systematically represented the results.

Table 3. Specificity results of five Malaria Pan Ag immunochromatographic rapid test kits formulated using different sizes AuNPs. Kit's specificity was tested using *P. Falciparum* positive blood specimens of concentration 40 parasites/µL. The color band intensity for control (C) line was +3 (high) and no line appears on the test zone (Pan) indicating absence of any cross reactivity.

Specificity of Kit Developed by 10 nm of AuNPs	Specificity of Kit Developed by 20 nm of AuNPs	Specificity of Kit Developed by 30 nm of AuNPs	Specificity of Kit Developed by 40 nm of AuNPs	Specificity of Kit Developed by 50 nm of AuNPs

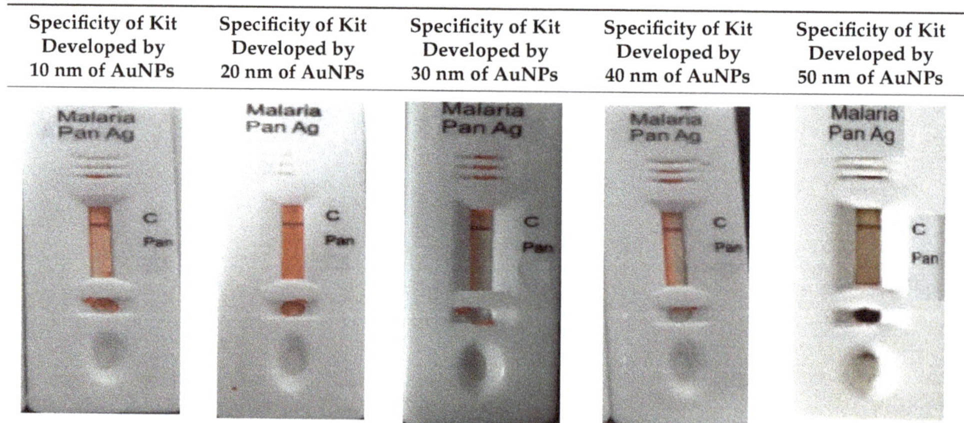

4. Conclusions

It can be concluded from the above study that the particle size of AuNPs has no effect on the test zone band intensity of Malaria Pan rapid diagnostic test kits if optical density of AuNP–protein conjugate is adjusted at 520 nm. The test zone band intensity was also observed to be maximum at 520 nm and optical density 3.0. It was found from the study that any quantity of protein can be utilized for AuNP conjugation, if the final optical density (OD) is adjusted correctly. It can also be concluded that, by optimizing the optical density, an enhanced test zone band intensity can be obtained while reducing the total number of trial and wastage of reagents.

Author Contributions: Conceptualization: P.M. and D.N.K.; Methodology: R.N and A.S.; Validation: A.K., J.D.S. and R.K.; Formal Analysis: A.F.V.L.; Investigation: S.J.B.F.; Data Curation: N.S.; Writing—Original Draft Preparation: P.M., R.N. and A.S.; Writing—Review and Editing: A.K.; Visualization: P.P.; Funding acquisition: D.O.B., S.K.P. and P.V.; Supervision: D.N.K. All authors have read and agreed to the published version of the manuscript.

Funding: This research received no external funding.

Conflicts of Interest: The authors declare no conflict of interest.

References

1. Mishra, M.N.; Misra, R.N. Immunochromatographic methods in malaria diagnosis. *Med. J. Armed Forces India* **2007**, *63*, 127–129. [CrossRef]
2. Manta, P.; Agarwal, S.; Singh, G.; Bhamrah, S.S. Formulation, development and sensitivity, specificity comparison of gold, platinum and silver nano particle based HIV $\frac{1}{2}$ and hCG IVD rapid test kit (Immune chromatoghraphic test device). *World J. Pharm. Sci.* **2015**, *4*, 1870–1905.
3. Maltha, J.; Gillet, P.; Jacobs, J. Malaria rapid diagnostic tests in endemic settings. *Clin. Microbes Infect.* **2013**, *19*, 399–407. [CrossRef] [PubMed]
4. Koczula, K.M.; Gallotta, A. Lateral flow assays. *Essays Biochem.* **2016**, *60*, 111–120. [CrossRef] [PubMed]
5. Gronowski, A.M. *Handbook of Clinical Laboratory Testing during Pregnancy*; Springer Science & Business Media: Berlin, Germany, 2004. [CrossRef]
6. Benjamin, G.Y.; Bartholomew, B.; Abdullahi, J.; Labaran, L.M. Prevalence of Plasmodium falciparum and Haemoglobin Genotype Distribution among Malaria Patients in Zaria, Kaduna State, Nigeria. *South Asian J. Parasitol.* **2019**, *23*, 1–7. [CrossRef]
7. Ramachandran, S.; Singhal, M.; McKenzie, K.G.; Osborn, J.L.; Arjyal, A.; Dongol, S.; Baker, S.G.; Basnyat, B.; Farrar, J.; Dolecek, C.; et al. A rapid, multiplexed, high-throughput flow-through membrane immunoassay: A convenient alternative to ELISA. *Diagnosis* **2013**, *3*, 244–260. [CrossRef] [PubMed]
8. Tsai, T.T.; Huang, T.H.; Chen, C.A.; Ho, N.Y.; Chou, Y.J.; Chen, C.F. Development a stacking pad design for enhancing the sensitivity of lateral flow immunoassay. *Sci. Rep.* **2018**, *8*, 1–10. [CrossRef] [PubMed]
9. Wong, R.; Tse, H. *Lateral Flow Immunoassay*; Springer Science & Business Media: Berlin, Germany, 2008. [CrossRef]
10. Lou, S.; Ye, J.Y.; Li, K.Q.; Wu, A. A gold nanoparticle-based immunochromatographic assay: The influence of nanoparticulate size. *Analyst* **2012**, *137*, 1174–1181. [CrossRef] [PubMed]
11. Kim, D.S.; Kim, Y.T.; Hong, S.B.; Kim, J.; Heo, N.S.; Lee, M.K.; Lee, S.J.; Kim, B.I.; Kim, I.S.; Huh, Y.S.; et al. Development of lateral flow assay based on size-controlled gold nanoparticles for detection of hepatitis B surface antigen. *Sensors* **2016**, *16*, 2154. [CrossRef] [PubMed]
12. Fang, C.; Chen, Z.; Li, L.; Xia, J. Barcode lateral flow immunochromatographic strip for prostate acid phosphatase determination. *J. Pharm. Biomed. Anal.* **2011**, *56*, 1035–1040. [CrossRef] [PubMed]
13. Safenkova, I.; Zherdev, A.; Dzantiev, B. Factors influencing the detection limit of the lateral-flow sandwich immunoassay: A case study with potato virus X. *Anal. Bioanal. Chem.* **2012**, *403*, 1595–1605. [CrossRef] [PubMed]
14. Banerjee, S.; Gautam, R.K.; Jaiswal, A.; Chattopadhyaya, M.C.; Sharma, Y.C. Rapid scavenging of methylene blue dye from a liquid phase by adsorption on alumina nanoparticles. *RSC Adv.* **2015**, *5*, 14425–14440. [CrossRef]
15. Kaur, K.; Forrest, J.A. Influence of particle size on the binding activity of proteins adsorbed onto gold nanoparticles. *Langmuir* **2012**, *28*, 2736–2744. [CrossRef]
16. Turkevich, J.; Stevenson, P.C.; Hillier, J. A Study of the nucleation and growth processes in the synthesis of colloidal gold. Discuss. *Faraday Soc.* **1951**, *11*, 55–75. [CrossRef]
17. Frens, G. Controlled Nucleation for the Regulation of the Particle Size in Monodisperse Gold Suspensions. *Nat. Phys. Sci.* **1973**, *241*, 20–22. [CrossRef]
18. Khlebtsov, B.N.; Tumskiy, R.S.; Burov, A.M.; Pylaev, T.E. Quantifying the Numbers of Gold Nanoparticles in the Test Zone of Lateral Flow Immunoassay Strips. *ACS Appl. Nano Mater.* **2019**, *2*, 5020–5028. [CrossRef]

19. Byzova, N.A.; Safenkova, I.V.; Slutskaya, E.S.; Zherdev, A.V.; Dzantiev, B.B. Less is more: A comparison of antibody–gold nanoparticle conjugates of different ratios. *Bioconjugate Chem.* **2017**, *28*, 2737–2746. [CrossRef] [PubMed]
20. Larm, N.E.; Essner, J.B.; Pokpas, K.; Canon, J.A.; Jahed, N.; Iwuoha, E.I.; Baker, G.A. Room-temperature turkevich method: Formation of gold nanoparticles at the speed of mixing using cyclic oxocarbon reducing agents. *J. Phys. Chem. C* **2018**, *122*, 5105–5118. [CrossRef]
21. Sujitha, M.V.; Kannan, S. Green synthesis of gold nanoparticles using Citrus fruits (Citrus limon, Citrus reticulata and Citrus sinensis) aqueous extract and its characterization. *Spectrochim. Acta Part A Mol. Biomol. Spectrosc.* **2013**, *102*, 15–23. [CrossRef]

MDPI
St. Alban-Anlage 66
4052 Basel
Switzerland
Tel. +41 61 683 77 34
Fax +41 61 302 89 18
www.mdpi.com

Diagnostics Editorial Office
E-mail: diagnostics@mdpi.com
www.mdpi.com/journal/diagnostics